Unlock genetics

in 10 concise chapters

Unlock

genetics

in 10 concise chapters

Marianne Taylor

This edition published in 2025 by Sirius Publishing, a division of Arcturus Publishing Limited,
26/27 Bickels Yard, 151–153 Bermondsey Street,
London SE1 3HA

Copyright © Arcturus Holdings Limited

All rights reserved. No part of this publication may be reproduced, stored in a retrieval system, or transmitted, in any form or by any means, electronic, mechanical, photocopying, recording or otherwise, without prior written permission in accordance with the provisions of the Copyright Act 1956 (as amended). Any person or persons who do any unauthorised act in relation to this publication may be liable to criminal prosecution and civil claims for damages.

ISBN: 978-1-3988-5772-8
AD011638UK

Printed in China

CONTENTS

Introduction
WHAT IS GENETICS? 6

Chapter 1
DNA 8

Chapter 2
CHROMOSOMES AND GENES 20

Chapter 3
PROTEIN SYNTHESIS AND SELF-REPLICATION 34

Chapter 4
ALLELES 46

Chapter 5
INHERITANCE 58

Chapter 6
EVOLUTION 70

Chapter 7
ARTIFICIAL SELECTION AND GENETIC ENGINEERING 82

Chapter 8
EPIGENETICS 92

Chapter 9
GENES AND HEALTH 100

Chapter 10
THE FUTURE OF GENETIC SCIENCE 114

Index 126

Credits 128

WHAT IS GENETICS?

IT'S IN THE GENES. What is in the genes? Well, nearly everything about who we are and what we do is influenced, if not outright dictated, by our genetic make-up. Not only that, but the genetics of the other organisms that are significant in our lives, from the viruses and bacteria that can cause disease, to the plants and animals we keep and farm, are increasingly well-understood, important and within our power to manipulate and change.

A BLUEPRINT FOR LIFE

Your genes came to you courtesy of your parents, plus a sprinkling of random changes that are unique to you. Looking back through time, you can envision an unbroken line of reproducing organisms – most recently the humans who were known to you, then more distant ancestors who you may know about from family stories, through more and more generations of people stretching back through the centuries and millennia. Keep going, and you will reach ancestors who were precursors to modern *Homo sapiens*, to early mammals busy dodging the footfalls of mighty dinosaurs, and on through early vertebrates to their invertebrate ancestors. Eventually you will reach a phase in Earth's history, spanning more than a billion years, where all life was single-celled and very like the bacteria that still overwhelmingly dominate life on Earth today. Before even that time, it is likely that genes existed, in the form of free-living and self-replicating miniature strands of DNA's sister molecule, RNA. From that moment in time to this one, an unbroken line of genetic reproduction over an unimaginable number of generations led to the mind-bogglingly complex and highly organized collection of organic molecules that is you.

Most of us have a personal curiosity about genetics, especially when we look at our closest relatives and wonder how our family traits come to be expressed and why we sometimes see variation and sometimes consistency. News stories we hear about gene therapy for hard-to-treat health conditions are inspiring and encouraging, but thoughts of consuming genetically modified food might be worrying. Those who are not science-minded might feel happy enough not to delve into the mysteries of quantum mechanics or astrophysics, but genetic science is rather closer to home and with elements that are potentially within our control, so we might well feel it's important to have a grounding in its basics.

ABOUT THIS BOOK

The ten chapters of this book each look at one aspect of genetic science, from the structure of genes and chromosomes and their organization in cells, to how traits are inherited, the role of genetics in evolution, applications of genetic science in healthcare, the role of environmental factors to explain why some genes are expressed while others remain 'asleep', and finally some thoughts on the future of genetic science. Using non-technical language, snippets of history and plenty of clear and straightforward diagrams, we set about unravelling the double helix of DNA and cracking the code of life.

KEY VOCABULARY	
DNA	The molecule that serves as a blueprint for building the proteins that make up a living creature.
Chromosomes	Strands of DNA, which have a set structure, and come in matching pairs or larger sets in higher organisms.
Genes	A piece of DNA that codes for one particular protein - often comes in variants (alleles) which function differently and may be dominant or recessive to each other.
Mutation	A biochemical change to DNA that causes its 'output' to alter.
Inheritance	The transfer of DNA from one generation to the next (including genes that are expressed and genes that aren't).
Epigenetics	How gene expression can be turned on or off by external factors.

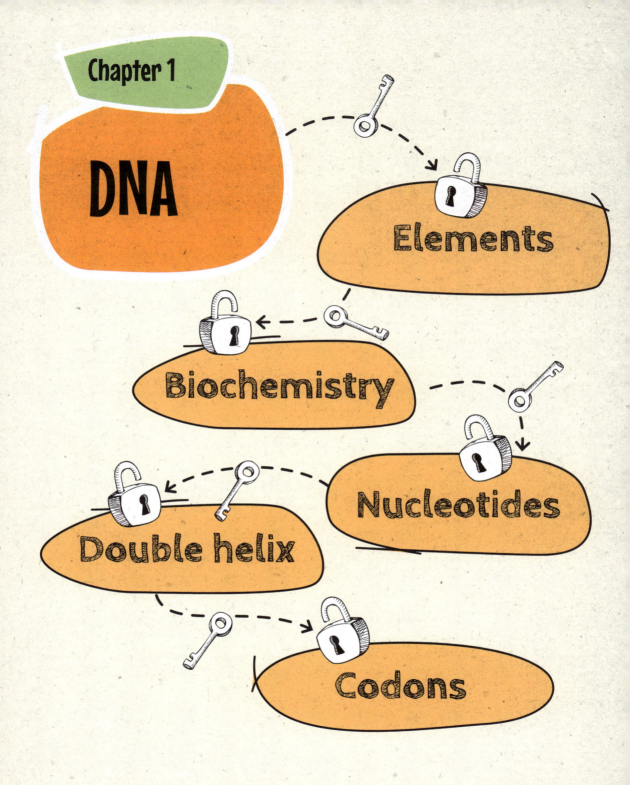

OUR BODIES, like all known matter in the universe, are composed of chemical elements. About 99 per cent of our mass is made up of just six elements – carbon, hydrogen, oxygen, nitrogen, calcium and phosphorus. Well over 50 more different elements make up the remaining 1 per cent. Some of these, such as sodium, chlorine, iron, iodine and copper, play vital roles in our bodily systems despite only being present in small or trace amounts, while others have no known function. The atoms of most elements are unstable in isolation so naturally form chemical bonds with other atoms (of the same or different elements) and exist as molecules. Our bodies contain thousands of different types of molecules, each with its own job to do, and the DNA molecule has a very important job when it comes to our genetics.

MOLECULAR MARVELS

The most abundant and one of the smallest molecules in our bodies is water (H2O) – each molecule comprises two hydrogen atoms bonded with one oxygen atom. At the other end of the scale is titin, short for 'titan protein'. Titin is a huge and complex, multi-folded protein found in muscle cells, and like other proteins it is made of carbon, hydrogen, oxygen and nitrogen. One titin molecule is composed of 34,350 amino acids, the building blocks of proteins (and each amino acid contains a minimum of ten atoms, so a titin molecule contains well over 350,000 atoms). DNA is short for deoxyribonucleic acid.

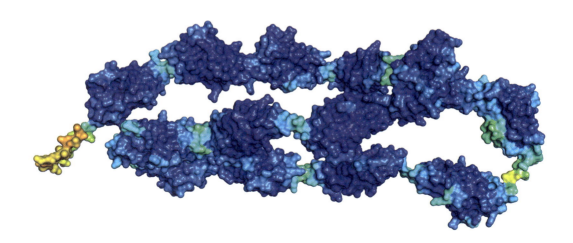

Model of a single titin molecule.

<< CHAPTER 1

You have probably also heard of RNA, or ribonucleic acid – this molecule interacts with DNA in various ways. These two molecule types are similar in structure and function, and collectively are called nucleic acids. They are polymers, meaning that they are made up of repeated 'building blocks' that join together in a chain, which can be short or long. Each individual building block or unit in a nucleic acid is called a nucleotide, and is made up of three parts: a pentose sugar, a phosphate group and a nitrogenous base. The key difference between DNA and RNA, in terms of the make-up of their nucleotides, is the type of sugar they contain – in RNA it is ribose, and in DNA it is deoxyribose.

DNA is often described as the body's genetic code or blueprint. All living things on Earth, from bacteria to Bactrian camels, have DNA in their cells and this wonder molecule contains a code that enables cells to put together amino acids in the right order to make every different type of protein that can be found in that particular organism's body. Each section of DNA that codes for a single specific type of protein is called a gene. In higher organisms the DNA is organized into several chromosomes, each of which contains multiple genes. The entirety of an organism's DNA is known as its genome. We inherit our DNA from our parents (50 per cent from each), and pass half of it on to each of our children – we also share a proportion of it with our other relatives.

The secret of how DNA molecules provide a genetic blueprint for building proteins is tied up to its chemical structure. The four different types of nucleotides (see opposite) that make up a strand of DNA can be chained together in any arrangement and repeated any number of times, like letters in different words, and the order of the nucleotide 'letters' tells the cells which proteins to build.

A family tree illustrates how we each inherit half of our DNA from our biological mother and the other half from our biological father.

DNA >> 11

First make your nucleotide

Let's take a look at the three components of a nucleotide, one by one.

Pentose sugar

Simple sugars or monosaccharides are small molecules made of carbon, oxygen and hydrogen, and have a ring structure. The best known example is glucose, which contains six carbon atoms and is a type of hexose sugar (*hexa* = Ancient Greek for six). Pentose sugars have five carbon atoms (*penta* = Ancient Greek for five). Four of those carbon atoms are held in the ring part of the molecule and the fifth in an 'offshoot' or 'tail'. The pentose sugar in RNA is ribose, which has the chemical formula $C_5H_{10}O_5$, while deoxyribose, found in DNA, has the formula $C_5H_{10}O_4$ – the 'deoxy' part indicates that it has one oxygen atom fewer than ribose:

Phosphate group

This component of the nucleotide comprises a phosphorus atom with associated oxygen atoms. It is bonded to the 'tail' of the pentose sugar molecule.

Nitrogenous base

The nitrogenous base or *nucleobase* is bonded to the other side of the pentose sugar molecule. It has either a single or a double ring-shaped structure and is made of a combination of nitrogen, hydrogen, carbon and oxygen atoms. In DNA, there are four types of nucleobases. They are adenine and guanine (which have a two-ring structure and are known as purines) and cytosine and thymine (which have single rings and are known as pyrimidines). In RNA, thymine is replaced by another pyrimidine, called uracil.

Ribose

Deoxyribose

The three pyramidine bases – thymine, cytosine and uracil.

The two purine bases – adenine and guanine

Because there are four different nitrogenous bases in DNA, there are four different types of nucleotides. The same goes for RNA. Within a chain of nucleotides, the four types can be in any order, and be repeated any number of times. At certain points in the strand, there are also attached methyl groups (see chapter 8 to learn how these chemical elements add a new dimension to the genetic landscape, through epigenetics).

A strand of RNA consists of a single chain of nucleotides. However, DNA is more complex. Its structure consists of not one but two chains of nucleotides, which are joined to each other via bonds that form between their nitrogenous bases, like the rungs of a ladder. A nucleotide on one side of the ladder can only pair with one type of nucleotide on the other side. The pairing is always a purine with a pyrimidine, with adenine always pairing with thymine, and guanine pairing with cytosine.

Comparison of RNA and DNA molecules.

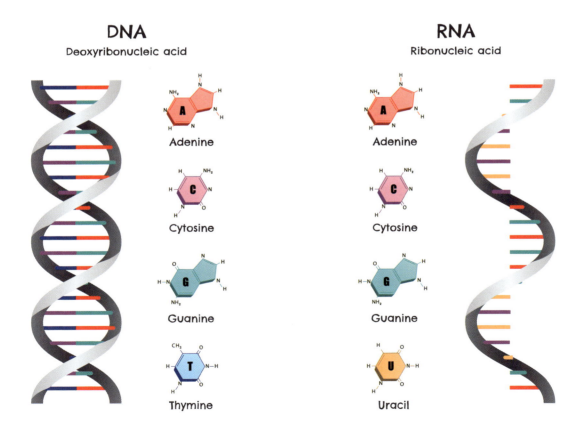

Each of the 20 types of amino acids that are present in the proteins in your body is coded by a specific sequence of three (triplet) particular nucleotides. These triplet sets of nucleotides are called codons. For example, the codon to make the amino acid tryptophan is thymine-guanine-guanine. A sequence of codons provides the sequence of amino acids needed to make a particular protein. There are also some codons that code for a 'stop now' signal, rather than an amino acid, signalling that the protein is now complete.

What about other molecules?

DNA is a code for making every type of protein in your body, but there are other kinds of large and complex molecules in your body besides proteins – they include lipids (fats), carbohydrates and more, including DNA itself. How does your body know how to make those? The many tasks our various proteins carry out include the construction of other types of molecules. *Enzymes* are proteins which are involved with *catalysing* (initiating or facilitating) chemical reactions to form or break chemical bonds in other molecules. In both cases, the 'input' is known as the substrate, and the 'output' as the product.

We may think of enzymes as mostly important for breaking down the food we eat into simpler molecules, and indeed we have many digestive enzymes that carry out this work. However, other enzymes are needed for molecules to be built. For example, digestive enzymes called amylases break down starch (a long polymer carbohydrate molecule) found in foods like potatoes. They turn the long starch molecules into simple glucose molecules, which are small enough to enter the bloodstream and travel into cells where they may be used immediately for energy. Some glucose molecules are joined back together but into a different polymer – glycogen rather than starch. This molecule is used as stored energy in our muscles and liver. The process of glycogen synthesis is initiated by the enzyme phosphorylase. In the synthesis of DNA, an enzyme called DNA polymerase is required to connect the sugar/phosphate elements of nucleotides together, and this is just one of many enzymes involved in constructing a DNA molecule.

How enzymes build molecules.

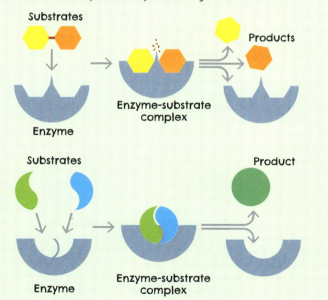

<< CHAPTER 1

DISCOVERING DNA

Humankind must have long observed that offspring can show some of the same traits as their parents – this is evident not only in our own children but in the animals we have bred for 6,000 years and the plants we have cultivated for about 23,000 years. However, we did not necessarily attribute this to some kind of biological inheritance, as we could also observe the effects of environmental factors on reproduction and development – for example, adverse weather could stop plants producing a good crop, and differences in their mothers' diet could influence the size and condition of newborn livestock. The book of Genesis in the Bible took this thought process further along (and very much down the wrong path, as we now know) in its description of how Jacob's sheep and goats were born spotted or striped if their parents had mated in front of spotted or striped tree branches.

The experiments of Gregor Mendel in the 19th century on pea plants (see chapter 5) showed that certain traits were shared between parents and offspring according to simple and predictable rules. This strongly suggested that there was a biological process of inheritance (rather than traits arising only from environmental factors), and scientists attempted to identify the substance that carried this coding from parent to offspring. The actual DNA molecule was isolated in 1868 by a Swiss medical student, Johann Friedrich Miescher, but its purpose of carrying the genetic code was not recognized for many more decades, as its structure was deemed too simple to account for the multiplicity of inherited traits. That DNA was in fact the molecule of heredity was discovered in 1952 by Alfred Hershey and Martha Chase, through their work on certain bacteriophage viruses. These viruses release their own DNA into bacterial cells, thus forcing the bacteria to make new virus particles.

Rosalind Franklin, studying the structure of DNA at King's College in London in the early 1950s, used X-rays to create images of the molecule, revealing

Below: (Left to right) James Watson (1928–), Francis Crick (1916–2004) and Rosalind Franklin (1920–58).

its helical structure. As she and her PhD student, Ray Gosling, prepared her findings for publication, her colleague Maurice Wilkins showed her photographs and interpretations to another pair of biologists, James Watson and Francis Crick. These Cambridge academics were trying to model the structure of DNA based on an understanding of how the nucleotides paired together, and Franklin's data enabled them to complete their model and publish a paper revealing the double helix, in the same 1953 issue of *Nature* in which Franklin's own data was published. Watson, Crick and Wilkins won the Nobel Prize in Physiology or Medicine in 1962 for determining the double helix structure of DNA, but Franklin's contribution was largely forgotten until recent times.

DNA IN CELLS

In our cells, DNA comes in two types and lives in two places. Nuclear DNA lives in the cell nucleus, and mitochondrial DNA or mtDNA lives inside our mitochondria, which are small structures or organelles found inside most of our cells. When a cell divides (mitosis), it copies its nuclear DNA with one copy going into each of the two daughter cells. Mitochondria divide independently of the cell in which they live, and make copies of their own DNA.

Some organisms reproduce asexually, effectively creating clones of themselves, meaning that offspring have the same DNA as their parent. When an organism reproduces sexually, it makes special sex cells (gametes) which carry only half of its DNA. When a male (sperm) and female (egg or ovum) gamete combine to form a single cell (zygote) that will develop into a new organism, that zygote has a full set of DNA made up of half of its mother's DNA and half of its father's. In both types of reproduction, DNA is inherited from one generation by the next, but in the case of sexual reproduction, the offspring's DNA is a unique mixture of its two parents' DNA. You can read more about this in chapter 3.

DNA IN THE NATURAL WORLD

As we have seen, all living things possess DNA. It goes even further than that though, as some viruses also have DNA, even though they are not technically classed as living things. Prokaryotes, which are single-celled living organisms with a simple cellular structure (they comprise the biological domains Archaea and Bacteria) have a single ring-shaped strand of DNA, the bacterial chromosome, which is folded up to fit inside the cell. Stretched out, the DNA of an *E. coli* bacterium is about 1.6mm long, greatly exceeding the length of the cell itself (about 0.004mm). Eukaryotes (organisms with complex cells – this includes plants, fungi and animals) have several strands of DNA, in the form of

The double helix

As we have seen, DNA is structured a bit like a ladder. Each supporting side is made up of the pentose sugar and phosphate group parts of the nucleotides, attached to each other in a continuous, supportive chain. Its rungs are formed by the paired nitrogenous bases of the nucleotide on each side.

In reality, DNA does not have a flat ladder shape. Imagine that your ladder is made not of rigid wood but of soft and flexible metal, which you can twist into a spiral shape without breaking any of its connections. This 3D spiral or double helix is the true shape of DNA, and in its natural state the helix itself is repeatedly coiled in an even more compact form.

The 'ladder' structure of DNA.

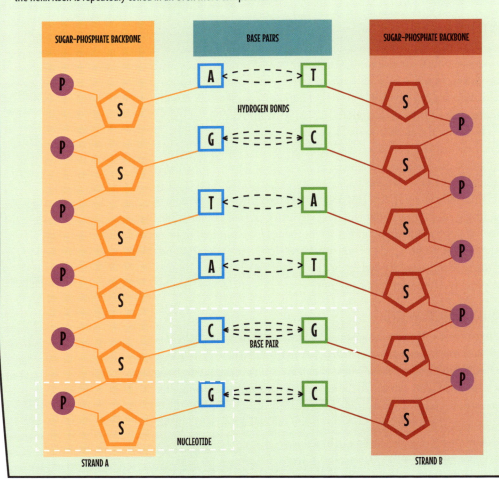

DNA >> 17

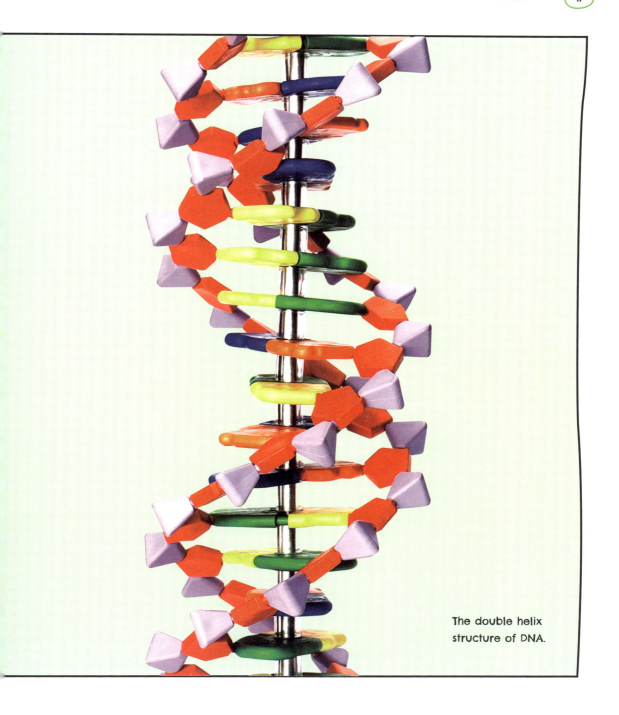

The double helix structure of DNA.

<< CHAPTER 1

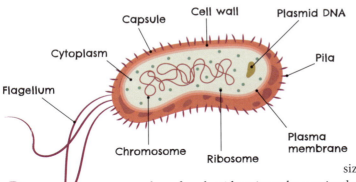

Diagram of an *E. coli* cell, showing its chromosome.

chromosomes. The DNA from one 0.01mm cell in your body would stretch out to about 2m (6.5ft), and if we made a single strand out of all of the DNA from all of your cells, it would span more than 18 billion kilometres (11 billion miles).

We usually consider the size of a genome in terms of the number of nucleotide pairs or base pairs that it contains. A human's entire complement of nuclear DNA, arranged in 23 pairs of chromosomes, contains about 3 billion base pairs, while that of the bacterium *E. coli* is about 4.6 million. However, species that don't seem that different to one another might have vastly different-sized genomes, and apparently complex species might have small genomes (and vice versa). The green-spotted pufferfish has 340 million base pairs, while the Australian lungfish has 43 billion. The largest known genome of any living thing belongs to a small fern-like plant native to Australia (there seems to be something about Australia and large genomes). *Tmesipteris oblanceolata* has 160 billion base pairs in its genome, despite being classified as a 'simple' plant that does not even produce flowers or seeds.

GENES AND JUNK?

As we have seen, one gene is the code for making one protein, and a protein is a string of amino acids in a particular order. One amino acid in a protein is coded for by a codon, which is a specific sequence of three base pairs within a strand of DNA. So how many base pairs make up a single gene? In the human genome, genes range from a few hundred base pairs to more than 2 million, but the average is about 1,000.

A large proportion of our DNA does not code for genes at all. Some of it has other known functions but other sections appear non-functional, including material left over from viral infections suffered by our ancient ancestors. This apparently non-functional DNA is sometimes known as junk DNA but this term is falling from favour as new research identifies possible roles that this material (or at least some of it) plays in the lives of our genomes.

The origins of mitochondria

The genome of a human mitochondrion contains just 16,569 base pairs. Rather than being separated into a number of separate chromosomes, mitochondrial DNA exists as a single ring-shaped strand. This may remind you of what we have just learned about the DNA of the bacterium *E. coli*, and that's no coincidence. Mitochondria began their existence as free-living bacteria, long before more complex life evolved. However, at some point in their evolutionary history, they became engulfed within other single-celled organisms and evolved into functional components of those other cells, in a process called *endosymbiosis*. This is why the mitochondria that exist today have their own separate DNA and reproduce independently of their 'parent' cell. It is a similar story with the plastids (various organelles such as chloroplasts) that exist within plant cells – all were originally free-living cyanobacteria, and now carry out the function of photosynthesis within their 'parent' cells.

When a sperm fertilizes an egg and a zygote is formed, all of that zygote's DNA-carrying mitochondria are contributed by the egg cell only. The much smaller and simpler sperm cell does contain some mitochondria but they lack mitochondrial DNA. This means that mitochondrial DNA is inherited strictly along the female lineage (matrilineal line) – sons and daughters both inherit it from their mother, but only daughters pass it on to their own children. When studying the family tree of a person, or the evolutionary history of a species, it is useful to look at both nuclear DNA and mitochondrial DNA, as both are passed on from parent to child through the generations.

The process of endosymbiosis.

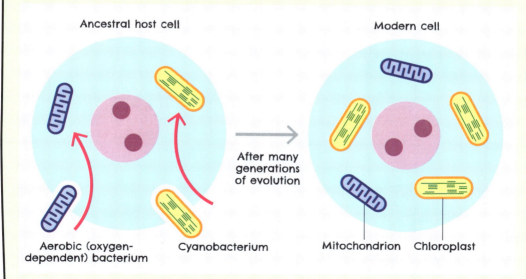

Chapter 2
Chromosomes and genes

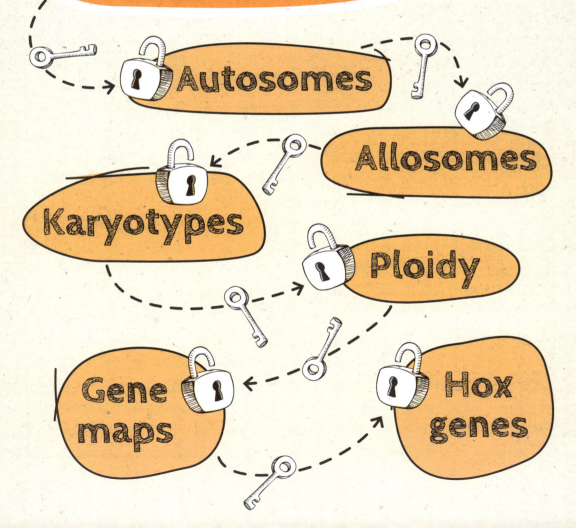

YOUR DNA EXISTS AS 46 CHROMOSOMES, in 23 pairs, each of them containing the coding instructions for making an array of different proteins. Chromosomes normally appear tangled together like a mass of bits of string inside a cell nucleus, but when the cell divides the chromosomes become shorter and thicker (condensed) and then duplicate themselves, and at this stage their structure is easier to visualize.

Because of chromosomes, a single fertilized egg 'knows' how to divide and divide again into a population of hundreds, thousands, millions and eventually trillions of cells, differentiated into about 400 different types, that make up an adult human. The exact number of chromosomes and their sizes shows tremendous variation across the natural world, although there are commonalities too.

46, MORE OR LESS

In humans, 22 of the 23 pairs of chromosomes are numbered from 1–22, with 1 the largest and 22 the smallest. These numbered chromosomes are known as autosomes, and the two members of each pair of autosomes are the same size and carry the same genes in the same order. The final pair of chromosomes are the sex chromosomes, or allosomes, which are designated by letters rather than numbers, and come in two different types (X and Y). With 22 pairs of autosomes and 1 pair of allosomes, this adds up to 23 pairs and a total chromosome count of 46.

The two types of allosomes that we have differ quite dramatically in size, appearance and the genes they carry, with X being much larger than Y. The system of designating autosomes with numbers and the single pair of allosomes with letters is used for the genomes of other organisms. The names and arrangement of the allosomes may vary. For example, the allosomes of mammals are X and Y, and in females the allosome pairing is XX, while in males it is XY. However, the allosomes of birds and some other organisms are known as Z and W rather than X and Y, but Z is still a lot larger than W, and in birds it is females that have the unmatched allosomes (male = ZZ and female = ZW). We refer to the sex with matching allosomes as the homogametic sex (meaning that all the gametes or sex cells they produce will contain the same sex chromosome). The sex with two different allosomes is the heterogametic sex (meaning that 50 per cent of their gametes carry one type of allosome, and the other 50 per cent carry the other).

<< CHAPTER 2

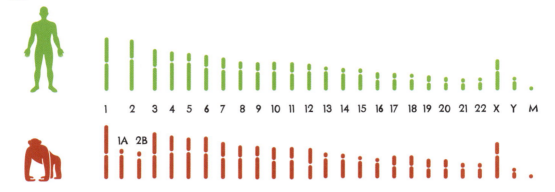

Human and chimpanzee genomes.

The genome of our closest living relative, the common chimpanzee, comprises 48 chromosomes (24 pairs), as do the genomes of the other living great apes. At some point in human evolutionary history, two chromosomes fused into one, giving us one pair fewer than our relatives. Studies of human and chimp genomes have revealed that the human chromosome 2 exists as two smaller separate chromosomes in chimps.

Humans and chimps are not the only closely related animals that differ in chromosome count, and the differences can be quite dramatic. The various species of cats (family Felidae) all have either 36 or 38 chromosomes (18 or 19 pairs), but the different dog species (Canidae) have between 36 and 78 chromosomes (18 to 39 pairs). As with the chimp and human example, differences in chromosome count in related species is usually down to fusion or splitting of chromosomes, rather than a big difference in the actual genes that make up those chromosomes. However, two species with different numbers of chromosomes are much less likely to be able to breed together and produce fertile offspring than two that share a chromosome count.

ANATOMY OF A CHROMOSOME

In its natural state, for most of the time a chromosome is a long, fine strand of DNA, wrapped tightly around proteins called histones which hold it in a stable

Colouring in

Those who know their Ancient Greek will have noticed that 'chromosome' means 'coloured body' (*khrôma* = colour, *sôma* = body). Chromosomes are not inherently colourful, though. The nature of chromosomes was first properly studied when biologists used chemical stains to highlight structures within cells while the cells were in the process of dividing (mitosis), and among the features that the stains revealed were the condensed chromosomes in the cell nuclei, caught in the act of replicating themselves and showing a distinctive pattern as different parts of them took up the stain in different ways. The name 'chromosome' was coined in 1888 by the German biologist Heinrich Wilhelm Gottfried von Waldeyer-Hartz. It took many more years for the structure and function of chromosomes to be fully elucidated.

CHROMOSOMES AND GENES >>

Of mules and hinnies

Horses and donkeys have both been domesticated for many centuries, and in their different ways they are both very useful animals for humans to have around as well as making intelligent and friendly companions. They are also clearly close cousins in terms of their evolutionary history. So it is no surprise that we have tried breeding them together many times, in the hope of combining their good qualities in a single animal. This cross-breeding is successful, in as much as a horse and a donkey can mate (as long as they are not too different in body size!) and this can lead to a viable pregnancy. The result of the mating is a hybrid horse x donkey which is called a mule if the parents were a male donkey and a female horse (the easier and therefore commoner pairing), and a hinny if the other way around. Today, nearly all mules and hinnies are produced via artificial insemination, so there is no practical need for the parent animals to be close in size (or to get along with each other or even meet!).

However, here the experiment must end, because both mules and hinnies are almost always sterile. The parent species, though similar, have different chromosome counts (64 in horses and 62 in donkeys), and mules end up with 63 – 32 from the horse parent and 31 from the donkey parent. Having one extra (or one missing) chromosome means that their bodies are unlikely to be able to make viable gametes (sex cells), because of the way chromosomes are divided up when gametes are formed (see chapter 3 for more about this). Nevertheless, people do still intentionally breed mules and hinnies for their own sake, and the name 'mule' is used for some other hybrids, for example the offspring of a domestic canary and a goldfinch or other wild finch, and the progeny of a mallard and a Muscovy duck.

Making a mule.

Female horse   Male donkey   Mule

Chromosomes = 64
Chromosome pairs = 32

Chromosomes = 62
Chromosome pairs = 31

Chromosomes = 63

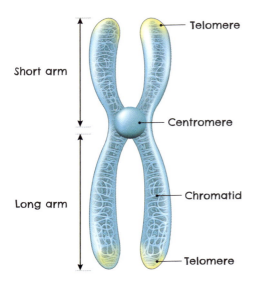

A chromosome, mid-division.

Different types of chromosome.

and compact shape. We now know that histones are important for more than providing structure and stability, and that they (together with methyl groups) comprise what is known as the epigenome (see chapter 8 for more on this).

When we picture a chromosome we tend to see an X shape, as on the left. This diagram shows a chromosome in its condensed form and in the process of duplicating itself, so really we are looking at two chromosomes, or sister chromatids. It is during this stage of cell division, when the two sister chromatids are still connected via the centromere (see chapter 3 for a description of the process of division) that it's easiest to visualize the structure of a chromosome. The centromere is a region in a chromosome that is important during replication, keeping the original and duplicate chromatids together as they reposition prior to cell division. Centromere position is only easily seen during chromosome replication, where it is the 'cross-over' point of the two joined chromatids in their X-shape. The position of a centromere is not perfectly central, giving the chromosome a short arm and a long arm, but the degree of asymmetry varies, with the short arm sometimes miniscule and sometimes almost as long as the long arm.

When treated with a stain, coloured bands appear in the chromosome (see Colouring In, p.22), marking out different regions. Some of these

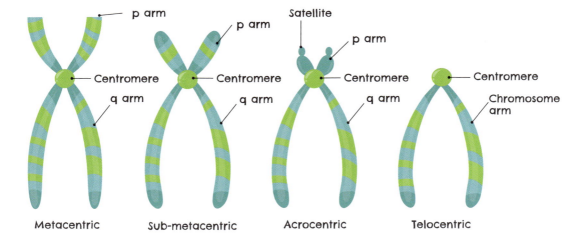

CHROMOSOMES AND GENES >>

cytogenetic bands are rich in either active DNA (euchromatin), others in less active or silent DNA (heterochromatin) and the two types stain differently when treated, making a striped pattern. The characteristic pattern of each chromosome helps us to describe the position of different genes within the chromosome (see Mapping Genes, p.29). The ends of a chromosome (telomeres) consist of non-coding DNA. They serve as a buffer zone, protecting the coding parts of the chromosome from damage, and tend to become shorter over the generations of cell replication. Telomere shortening is associated with signs of physical aging, although interestingly in a few organisms (such as storm petrels, which are small but very long-lived seabirds) telomeres have been found to actually lengthen with age.

KARYOTYPING

A depiction of a person or other organism's full set of chromosomes is known as a karyotype. Typically the chromosomes are shown in a standardized way, arranged tidily in their chromatid pairs in a vertical alignment, with long arms pointing down and short arms pointing up, and in a large to small size order to match the numbering system used for that species, in an image called a karyogram. This is usually a composite of photomicrographic images,

Human karyogram showing Down syndrome (three copies of chromosome 21).

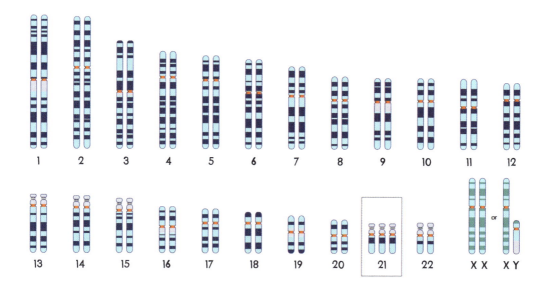

but biologists may also create a graphical representation of the chromosomes based on such photos.

A karyogram reveals at a glance whether an individual's chromosomes are typical for their species. Karyotyping will also show the genetic sex. If a human has a trisomy condition caused by an additional chromatid alongside the expected pair, such as Patau syndrome which is caused by an extra copy of chromosome 13, this will be evident from their karyogram, and other significant abnormalities of chromosomes will also be apparent. Karyotyping provides a 'big picture' of chromosomal structure and is a starting point when it comes to diagnosing a suspected genetic anomaly, but the detail may be insufficient to discern, for example, simple mutations in single genes. However, as technology progresses, more and more detail can be discerned from a karyogram.

Redrawing of human karyogram.

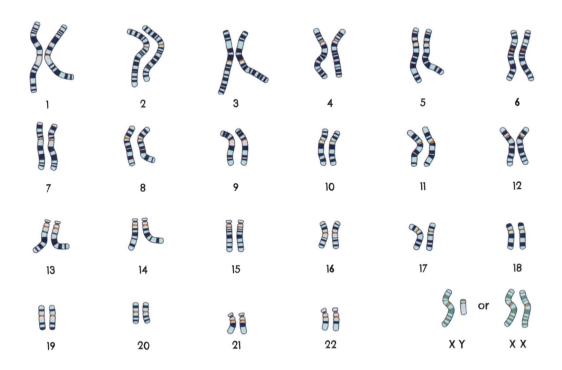

CREATING A KARYOGRAM

Photomicrographic images to be used for karyotyping need to take sample material from cultured cells that are in the process of dividing – specifically from the metaphase or prometaphase of the mitotic cell division cycle, when the chromosomes have replicated and are in their most condensed form. The cells are treated with toxic chemicals to freeze the chromosomes in this form, and then with a hypotonic solution (less concentrated than the fluid within the cell). Through osmosis, water molecules enter the cell from the surrounding solution, and cause it to burst. The chromosomes can then be extracted, treated with staining dyes to show contrasting banding patterns (these help identify which chromosome is which as well as reveal abnormalities), and photographed via a microscope. A composite image can then be created from multiple photos, to show all the chromosomes in their pairs.

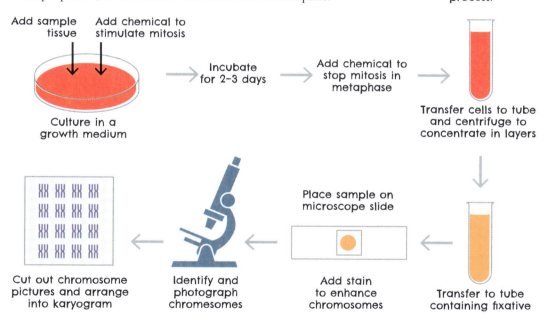

Stages of the karyotyping process.

PLOIDY

The word ploidy refers to the number of sets of chromosomes an organism has in its cells. We have two sets (i.e. our chromosomes come in pairs) in the majority of our cells and this makes us diploid. Before fertilization, both of the gametes involved (sperm and egg) contain only one set of chromosomes (23 individual chromosomes in each). These cells are known as haploid, because it is through a special process of cell division, meiosis, happening only in gametes, that they have half the number of chromosomes than is typical for a cell of their species. When they fuse to make a single cell, that cell is a diploid zygote because the egg's 23 chromosomes plus the sperm's go together to make a full diploid set. Most types of bacteria have monoploid cells because they have just one chromosome, and the same is true for the mitochondria in our cells (see p.17). Male honeybees are also monoploid, which means that a queen honeybee does not need to mate in order to produce male offspring – the unfertilized eggs she lays will result in male offspring.

Any cell with three or more sets of chromosomes is known as polyploid. We tend to think of this as an atypical state, because in mammals (and most other vertebrates), polyploidy is usually the result of a problem during the formation of gametes or during fertilization, and even if a polyploid zygote does form and begin to develop, it is very likely to die at an early stage. However, polyploidy is extremely common in some other groups of organisms. Many plants are polyploid, perhaps as many as 80 per cent of all

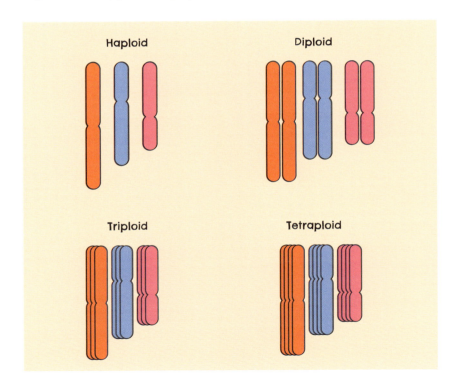

Common types of ploidy in living organisms.

species of flowering plants. Many of these have four sets of chromosomes in their cells (tetraploid), but some have as many as 64 (tetrahexacontaploid). Polyploidy is also fairly common in some animal groups, and even some of the cell types in our own bodies are polyploid – for example, some liver cells are tetraploid and some are octaploid (eight chromosome sets). If an organism has an unbalanced number of chromosomes, for example one of the chromosomes in just one pair is missing, or duplicated, this is called aneuploidy.

MAPPING GENES

In 1976, the very first full mapping of a genome was completed. The genome in question belonged to a bacteria-attacking RNA virus known as MS2. It was a tiny genome, coding for only four genes, but this paved the way for exploration of larger genomes, including that of another more complex virus (bacteriophage λ, which has DNA rather than RNA) with about 100 genes in 1982. In 1995 the first genome sequence for a free-living bacterium was completed – the species in question was *Haemophilus influenzae*, which has more than 1,600 protein-coding genes and is responsible for a variety of unpleasant infections, from meningitis to pneumonia. Meanwhile, something rather more ambitious had already begun. In October 1990, the Human Genome project was launched. In April 2003, to much fanfare in the scientific and wider community, it was completed, and stands today as one of the great advancements of modern science – a testament to the hundreds of researchers who put in thousands of laboratory hours to crack the human DNA code. Its findings have revolutionized practical healthcare as well as our understanding of human biology and evolution, and today the genomes of many more organisms have been fully mapped.

Thanks to this project, the locations and functions of all protein-coding genes in humans are now known.

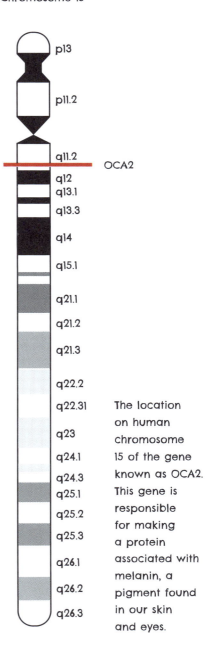

The location on human chromosome 15 of the gene known as OCA2. This gene is responsible for making a protein associated with melanin, a pigment found in our skin and eyes.

The position of a gene on a chromosome is called its locus. Each locus has an address which is made up of three parts – the chromosome number, whether it is on the long or short arm of the chromosome (designated with q and p respectively), and its position on the arm, in relation to the sub-bands (cytogenic bands) that are visible when the chromosome is stained. For example, the OCA2 gene, mutations of which can cause albinism (the condition of having no melanin pigment in cells, and so having white skin and hair, and pink eyes) in humans and many other mammals, has the locus 15q11-q13 in humans. This means it resides on our chromosome 15 on the long (q) arm, in the region of cytogenic bands 11–13.

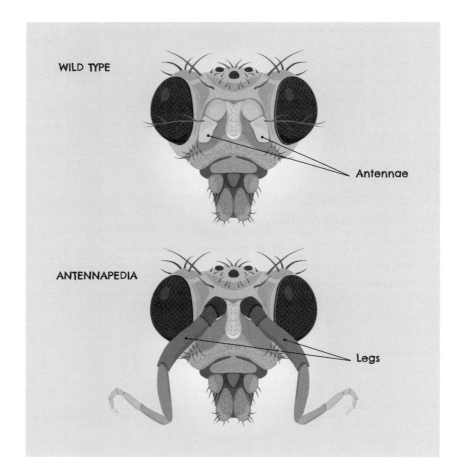

The fruit fly variation 'antennapedia', in which the fly grows legs where its antennae should be, because of a hox gene mutation.

CHROMOSOMES AND GENES >> 31

HOX GENES

Fruit flies (*Drosophilia*) might not be welcome in your kitchen, but they are popular laboratory animals, because they are easy to care for and breed quickly. The study of genetics owes much to these little insects, and one of the best known discoveries came about when studying *Drosophilia* with the condition known as antennapedia – instead of a pair of antennae on their heads, these individuals have a pair of legs.

That such a large-scale abnormality could be caused by a mutation in just one gene seemed unlikely, but led to the discovery of a special class of genes known as hox genes. These genes are active during early embryonic stages of animals with bilateral symmetry (body parts mirrored on each side), and regulate patterns of development of major anatomical structures. The hox gene initiates the process of building a leg, or an antenna or other body part, in its correct anatomical place, and a mutation in this gene can mean that the structure is built in the wrong place, or built too many or not enough times. In humans there are 39 different hox genes, and several known related disorders, including polydactyly (having more than the usual number of fingers and/or toes).

THE SELFISH GENE

Richard Dawkins' book *The Selfish Gene*, published in 1976, introduced the idea that genes are in competition for survival. When we act in a way that preserves our lives or favours our health and survival, we might be considered to be selfish, but what about when we act altruistically, setting aside self-interest for the good of our family? Acts of altruism might seem anomalous when evolutionary ideas might tell us that survival and procreation is every individual's overarching goal. We can't put altruistic acts down to fine and noble human sensibilities either, since altruism is readily observed in many other species too. That animals will sacrifice themselves for their own young while at the same time fight fiercely with other unrelated individuals of their species for resources, suggests that altruistic acts are carried out primarily not for 'the good of the species' but for 'the good of the family'. Dawkins suggested that this 'kin selection' occurs because of genetic relationships – we want our genes to survive down the generations, so we act to further the survival of others who share our genes, as well as ourselves.

This is why a male lion, on taking over a pride, kills the young cubs – but later when he has fathered cubs in the pride himself, he treats these new cubs with care and affection. He is motivated not to act in the best interests

A male lion is an indulgent and patient parent, as long as he is sure the cub is his own.

of lions in general, but in the best interests of his own particular genes. The new cubs carry those genes, while the original cubs did not and would have been competitors for the resources that his own cubs needed. The selfish gene theory also explains why, in some animal species, social groups form in which all the breeding is done by only one dominant pair (or one dominant female with multiple males). The others in the group are related to the breeding individuals and therefore to all the youngsters born to them. When a dominant breeding animal dies, a subordinate may take over, but most subordinates only ever act as helpers. However, in doing so they are still putting their efforts into ensuring that their genes are passed on, via their close relatives. We do see altruism between unrelated individual animals too, but that tends to come with an expectation of a reciprocal act somewhere down the line.

Genes and memes

The term meme is most often used today to describe an image or snippet of text that is widely shared on the internet, but it can refer to any cultural idea or behaviour. It was coined by Richard Dawkins in *The Selfish Gene*. He saw the idea of a meme as being related to a gene, in that it is an entity that is passed from one mind to another, in the same way that a gene is passed from one generation to another. Some memes, like some genes, are destined to die out quickly while others will survive, thrive and spread very widely – though changes in the cultural climate could quickly kill off a once successful meme. Memes, like genes, may also mutate (pick up changes as they are passed along), and the mutant versions may be more successful (or less so) than the original. Just as genes are important in biological evolution, so memes are important in cultural evolution – a process which happens in a variety of social animals as well as in humans.

The 'ice bucket challenge' is an example of a 'real-life' (as opposed to digitally shared) meme. A charitable fundraising activity, it involved friends challenging one other to douse themselves in icy water and collect donations. The challenge went viral in the summer of 2014 and raised more than $115 million for various causes.

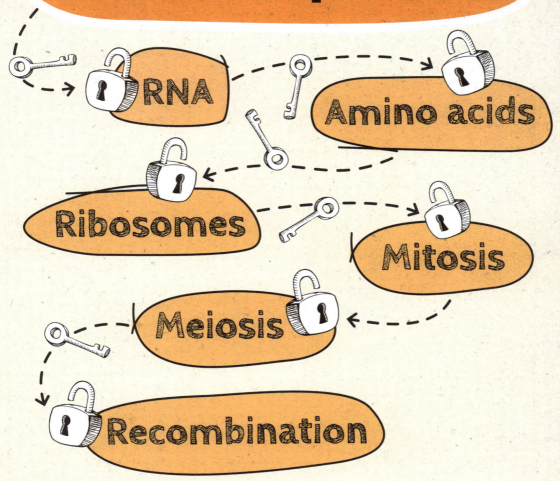

Chapter 3

PROTEIN SYNTHESIS AND SELF-REPLICATION ›› 35

THE PROCESS OF TRANSLATING the DNA code into actual proteins that the body can then use in a variety of ways is complex. It involves biochemical processes that are managed by a range of molecule types and take place in a variety of cell structures. The process of replicating DNA when a cell divides from one into two has some features in common with protein synthesis – and in both cases it is DNA's sister molecule, RNA, that plays a key part.

ABOUT RNA

We have already learned the basics about RNA, but let's recap. It is a nucleic acid found in living cells (and also in most viruses) and, like DNA, it consists of a long chain of repeated units, which are called nucleotides and are made of three parts – a pentose sugar, a phosphate group and a nitrogenous nucleobase. In DNA, the sugar is deoxyribose, and in RNA it is ribose. As with DNA, the nucleobases in RNA come in four types. Three of those types are the same as three of DNA's nucleobases – adenine, cytosine and guanine. The fourth is different – where DNA has thymine, RNA substitutes thymine with uracil. The third key difference is that DNA exists as a double strand, and RNA as a single strand.

The small size of RNA as a molecule means it can move into and out of the nucleus (and other organelles) of a eukaryotic cell, passing through pores in the nuclear membrane and moving through the cytoplasm that fills the rest of the cell. It can therefore carry the DNA coding instructions to where they are needed. Three main types of RNA are involved in the process of using DNA instructions to build a protein – they are messenger RNA (mRNA), transcription RNA (tRNA) and ribosomal RNA (rRNA).

Model of messenger RNA.

RNA world

Many scientists believe that a precursor to life on Earth must have been some kind of molecule that could replicate itself. This molecule could then have proliferated, and would have been subject to natural selection — errors in replication would have introduced variation in its populations, and some of those variants would have been more successful at surviving and replicating than others, leading to adaptation over time with the 'best' variants becoming better over successive generations. RNA is the most obvious candidate for this role, and all of its components would have existed on an early, pre-life Earth. However, in modern organisms RNA needs other molecules (including DNA) to carry out self-replication. In 2009, scientists at the Scripps Research Institute in California published their research on creating infinitely self-replicating RNA molecules in the lab, in the absence of DNA and other cellular entities. The discovery that this is possible supports the concept of an 'RNA world' picture of pre-life Earth.

How a theoretical 'RNA World' could have evolved, and led to the emergence of DNA and the first living organisms.

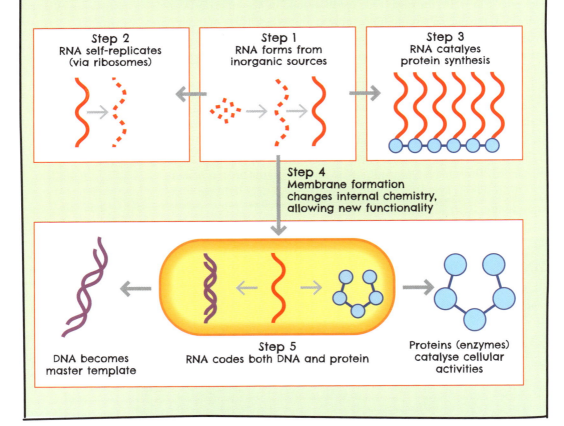

PROTEIN SYNTHESIS AND SELF-REPLICATION >>

PROTEIN SYNTHESIS

To make a protein, you need the gene that gives the code for the amino acids that are chained together to create that protein. You will find that gene on a strand of DNA. As we have seen, DNA resides in the cell nucleus in the form of chromosomes, but proteins are built by organelles called ribosomes, which exist in the cell cytoplasm, outside of the nucleus. This is where mRNA comes in. Its job is to be a template for that piece of genetic code, and to carry that code out of the nucleus, in order for it to bind to a ribosome where it will be 'read' and the protein built.

The process of making a piece of mRNA takes place in the cell nucleus, and is called transcription. An enzyme called RNA polymerase II is needed for this process, to assemble the NTPs (nucleoside triphosphates) – molecules that are precursors to nucleotides – in the correct order. The DNA strand is partly 'unzipped' (imagine the rungs of a ladder breaking down the middle) so that the NTPs of the new mRNA strand can be paired with the nucleotides of one half of the DNA strand. The strand that is used to form the mRNA is known as the template strand, while the corresponding strand (of which the new mRNA is a copy) is the coding strand.

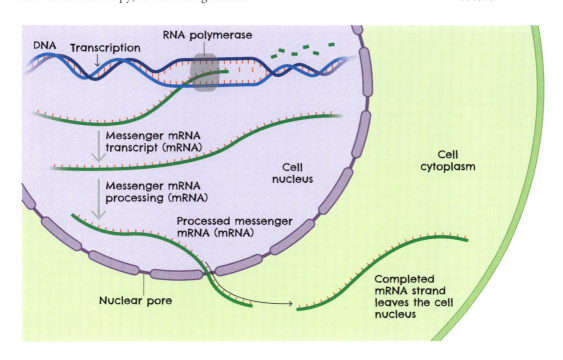

How a strand of mRNA forms, and carries the DNA's code out of the cell. nucleus

38 << CHAPTER 3

> ### Naming an enzyme
> Many enzymes are involved in the cellular processes around protein synthesis and DNA and RNA replication (and in all cellular processes generally). You can tell if a named molecule type is an enzyme by whether it ends in 'ase'. The first part of the name describes the substrate on which the enzyme acts, and the 'ase' denotes that it is an enzyme. For example, a lipase is an enzyme that breaks down lipids (fats) and a protease breaks down proteins. RNA polymerase is an enzyme that builds the RNA polymer (a polymer being any molecule that is a chain of repeated similar or identical units). There are multiple types of RNA polymerase, carrying out different variants of RNA transcription. RNA polymerase II is the variant involved in transcribing DNA coding into mRNA, while RNA polymerase I is involved in making rRNA, and RNA polymerase III primarily helps to make tRNA.

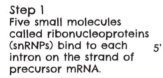

The stages of RNA splicing.

Step 1
Five small molecules called ribonucleoproteins (snRNPs) bind to each intron on the strand of precursor mRNA.

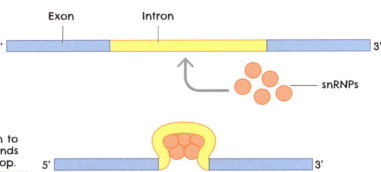

Step 2
The snRNPs cause the intron to fold in itself, bringing the ends close together to form a loop. The ends of the exons also move closer together to help close the loop.

Step 3
The intron now detaches and the two splice sites connect. With all introns removed, the strand becomes mature mRNA. The snRNPs detach from the intron and are used for more splicing, while the detached introns are usually used in other processes.

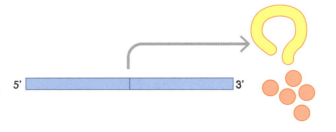

The new strand of mRNA still includes non-coding sections (introns) as well as coding sections (exons). At this stage it is known as precursor mRNA, and in order to become mature mRNA the non-coding introns need to be removed. It undergoes further processing within the cell nucleus, a process called RNA splicing wherein the introns are removed and the strand reattached at the break points. Also, the exons are often somewhat rearranged in a process called alternative splicing, meaning that the same gene can generate different mRNA strands which can code for different versions or isoforms of that gene's protein molecule. These isoforms may work in slightly different ways within cells. The splicing process is controlled by large, complex molecules of RNA combined with protein, called ribonucleoproteins or spliceosomes.

Once this has been done, the mature mRNA strand is ready to leave the nucleus. It exits via a nuclear pore and enters the cell cytoplasm, where a ribosome attaches to it. The next part of the process, translation, is ready to begin now – this is where the code that the mRNA has carried from the cell's DNA is read and from it a protein is built.

What's a ribosome?

Ribosomes are organelles found in cells, and are one of the few organelle types that exist in prokaryotic as well eukaryotic cells. They are often drawn as two joined blobs, one larger than the other, like a snowman or cottage loaf. This represents the two subunits from which they are made – the large subunit and the small subunit. Both parts are made of a combination of ribosomal RNA and proteins. The subunits of ribosomes are built in the nucleolus, a part of the cell nucleus. They enter the cytoplasm via the pores in the nuclear membrane, and are assembled after that. Some ribosomes exist freely in the cytoplasm, while others are bound to the membrane of another cellular structure called *endoplasmic reticulum*. Both free and bound ribosomes make proteins, although those made by free ribosomes generally have a function within the cell, while those made by bound ribosomes are more likely to leave the cell to carry out their function elsewhere.

Note: *When we think about translation, it's important to remember that the molecular machinery works on codons – the specific triplet sets of nucleotides that correspond to a particular amino acid, or a 'stop here' message. You could think of nucleotides as letters, and codons as three-letter words – the ribosome reads the code in mRNA as words, rather than individual letters.*

Assembling amino acids (Trp = tryptophan, Lys = lysine, Asp = aspartic acid, Phe = phenylalanine) in the order coded on the mRNA strand.

When it builds a protein, the ribosome forms a chain of amino acids based on the mRNA. The process of translation involves the use of tRNA, which comes in small molecules that are shaped like clover leaves. The central 'leaf' of the tRNA molecule is called an anticodon. It consists of a sequence of three nucleotides, corresponding to the inverse sequence of three on a molecule of mRNA. For example, a molecule of tRNA with the anticodon GCC (guanine, cytosine, cytosine) corresponds to the CCG on mRNA. At the 'stem' end of the tRNA molecule is a binding site for an amino acid – in our example, the amino acid that will bind here is alanine, as this is the amino acid that is coded by CCG. The anticodon AUG on tRNA (adenine, uracil, guanine) corresponds to UAC (uracil, adenine, cytosine) on mRNA. When the ribosome reaches a 'stop' codon it has completed the protein molecule, which is then detached.

Note: *UAC is the mRNA's version of the codon TAC in DNA – remember that uracil is the substitute for thymine in RNA. The codon TAC codes for the amino acid tyrosine.*

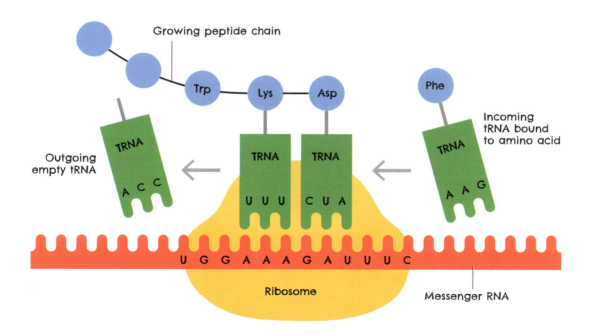

Joining the fold

Proteins need more than the right sequence of amino acids in order to be functional. Each protein molecule also has a distinct and stable three-dimensional shape, and if this is incorrect (misfolded) the protein will not function as it should, and in fact could even cause harm to the organism. Prions, which cause brain diseases such as bovine spongiform encephalopathy (BSE or 'mad cow disease') are misfolded variants of normal protein molecules. The folding stage of the process mostly takes place after the ribosome has completed assembly of the amino acid chain, and begins with parts of the chain naturally forming spirals or helices as chemical bonds form between different parts of the molecule. These are then folded and bonded again to create the three-dimensional form.

DNA REPLICATION IN MITOSIS

When a cell divides into two daughter cells, this process is called mitosis. Most types of somatic cell (non-sex cells) undergo repeated rounds of mitosis, although many have an 'end form' which is not able to divide again, for example our neurons or nerve cells. The cells in some types of tissues undergo continual mitosis, to grow or to replace cells that are worn out and not functioning correctly.

Prior to mitosis, the cell grows larger and duplicates many of its organelles, so there is enough material to turn it into two cells. This stage, known as interphase, is the longest, taking up about 90 per cent of the cell's lifetime.

The first stage of mitosis is called prophase. It involves all of the chromosomes in the cell's nucleus being duplicated, to create an exact copy of each chromosome. The process of duplication requires a quantity of free nucleotides to be present, and the use of a primer – a short chain of nucleotides to which new nucleotides are added according to the sequence of the original or template strand of DNA.

Now, the original DNA strand is unzipped and free nucleotides are added to the primer in an order corresponding to the sequence of the original DNA strand, facilitated by the enzyme DNA polymerase. The process occurs simultaneously at multiple points and on both sides of the original double helix, meaning that the new chromosome is built in fragments, which are then connected with the help of another enzyme, ligase. The two completed strands then twist together into the double helix shape, forming a new chromosome which is identical to its template. The two chromosomes, original and duplicate, remain connected at the centromere (see chapter 2) and are known as sister chromatids. At this point, a human cell contains 92 chromatids, connected as 46 pairs. All of the chromosomes condense down, assuming the familiar 'X' shape (which is visible under a microscope, with appropriate cell staining, and is used in karyotyping).

Next comes metaphase, in which the nuclear membrane breaks down, and the chromosomes line up across the centre of the cell (the point where it will be dividing into two cells). Now, in anaphase, the sister chromatid pairs are pulled apart, with one member of each pair moving into each half of the cell. In telophase, a new nuclear membrane forms around the two separated sets of chromosomes and in the final stage, cytokinesis, the cell itself completes its division as its plasma membrane closes around the central pinch point. Now two daughter cells exist, each with a full set of chromosomes. That is 46 in human cells, comprising 23 pairs.

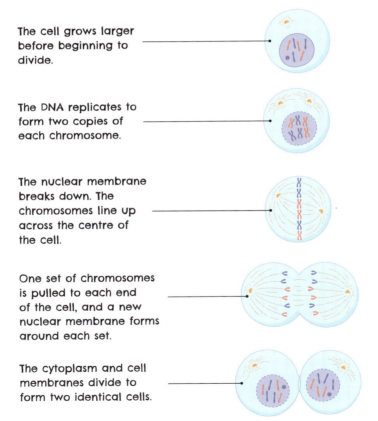

The process of mitosis – a parent cell divides into two genetically identical daughter cells.

The cell grows larger before beginning to divide.

The DNA replicates to form two copies of each chromosome.

The nuclear membrane breaks down. The chromosomes line up across the centre of the cell.

One set of chromosomes is pulled to each end of the cell, and a new nuclear membrane forms around each set.

The cytoplasm and cell membranes divide to form two identical cells.

DNA REPLICATION IN MEIOSIS – GENES FOR A NEW GENERATION

Meiosis is the process of cell division that results in mature gametes. During formation of gametes (gametogenesis), the precursor cells divide through normal mitosis several times. However, because gametes must contain only a single set of chromosomes (23 in the case of humans) rather than a paired set, the final part requires a second round of cell division, resulting in four daughter cells from one original cell, each daughter cell holding just 23 chromosomes.

Another crucial difference is that just after the chromosomes are duplicated in prophase, in meiosis they also undergo recombination. This means that sections of the DNA in the sister chromatids of each pair are exchanged with their opposite number. Remember that chromosomes come in pairs. Let's look at just one of these pairs within the nucleus of a cell – say, the two versions of chromosome 4. These two chromosome 4s have the same genes but may have different versions or alleles (see chapter 4 to learn about alleles) of many of those genes. The two chromosome 4s are called homologues. When this cell divides, each of the two chromosome 4s is duplicated into two sister chromatids, which start out being identical. In recombination, some of the DNA making up the sister chromatids of the first chromosome 4 is swapped with some of the DNA making up the sister chromatids in the second chromosome 4. This might be rather difficult to get your head around, but the diagrams will help (we hope).

How chromosomes exchange genetic material during recombination.

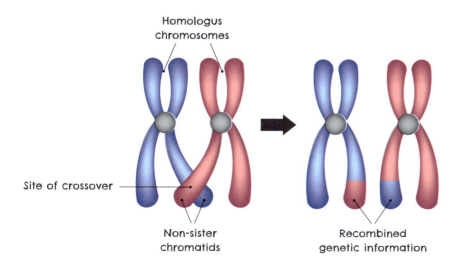

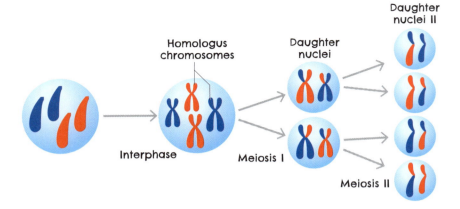

In meiosis, recombination ensures that each of the four daughter cells is genetically different.

The shuffling of DNA that happens in recombination, a process involving multiple enzymes, means that the sister chromatids, although not yet separated, are no longer twins, as they no longer have the same alleles. This is an important distinction from how chromosomes divide in mitosis. After recombination is complete, the sister chromatids then separate. Now the cell as a whole completes its division, in the same way as described for mitosis. These two daughter cells then divide one more time, resulting in four 'grand-daughter cells'. However, the second cell division happens without duplication of the chromosomes. So this time only one chromosome of each pair is separated into each of the four grand-daughter cells, and because of recombination, that single chromosome's DNA consists of a mix of the alleles present on both of the copies of that chromosome in the original cell.

Recombination means that no two gametes carry the same combination of alleles on their chromosomes, and no two zygotes formed from gametes from the same two parents will have the same combination of alleles. (Well, this is theoretically possible, by chance, but the odds against it are truly astronomical.) The exception to this is monozygotic twins, in which two embryos develop from a single zygote. Because they derive from the same egg and sperm, they have the same DNA. Monozygotic twins are rare in most mammal species, happening accidentally in early embryonic development, and the phenomenon is almost unknown in other vertebrates because that would require them to develop inside the same egg – physically very difficult, even if their mother did manage to lay an egg large enough to contain two young at full development.

Note: *Armadillos of the genus* Dasypus *are highly unusual in that most species routinely have monozygotic multiple births. One egg is fertilized, but divides into four embryos, so all litters are of identical quadruplets.*

When things go wrong

Making a copy of DNA, whether in mitosis or meiosis, is not error-free. It is extremely accurate, but it is also dealing with a colossal quantity of biochemical data – 3,200,000,000 different nucleotides in an entire genome. Many different kinds of copying mistakes or mutations can and do happen, with varied consequences. Mutations can happen in mitochondrial DNA as well as in nuclear DNA, and these also carry the possibility of significant consequences, given the importance of mitochondria in the function of cells and organisms.

In mitosis, if your immune system recognizes a new cell as abnormal (or dead) because of a significant mistake during the DNA replication process, that cell will probably be quickly removed. In meiosis, fertilization is likely to be impossible with seriously abnormal gametes. However, sometimes mistakes are not so damaging, and affected cells may continue to function and (in the case of many types of somatic cell) to proliferate. Some mutations in somatic cells can result in cancer, while mutations that occur during gametogenesis will be passed on to offspring if the affected gamete unites with a corresponding gamete and forms a zygote. Mutations can be harmful in the extreme, but they are also the source of variation in a population of an organism, creating new alleles which result in new protein types and, potentially, new traits. Without genetic mutation to create genetic variety, there would never have been any natural selection – and therefore no evolution. We look at particular types of mutations in chapter 4.

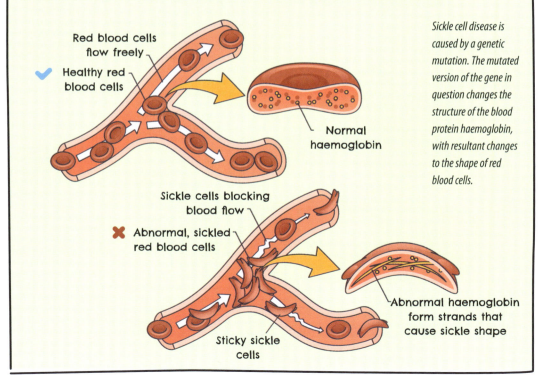

Sickle cell disease is caused by a genetic mutation. The mutated version of the gene in question changes the structure of the blood protein haemoglobin, with resultant changes to the shape of red blood cells.

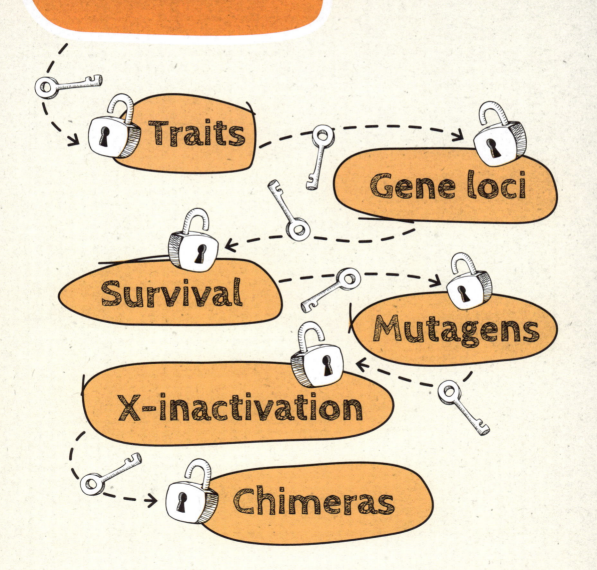

As we have seen, a gene is the code for making a protein, and our bodies use a great variety of different proteins in all kinds of cellular processes. However, we are all different, in both our outward appearance and how our bodies work. Even if you compare yourself to a close relative of the same sex, you will instantly think of dozens of life-long biological differences that probably can't be put down to different life experiences, from exact eye colour to differences in how much you enjoy and how well you digest different foods. We have the same genes, but how those genes (or rather the proteins they make) work in our bodies is highly variable and this variation is down to our different alleles.

GENES AND ALLELES

We often talk of 'having a gene for [X trait]'. In the case of a trait which is known or believed to be associated with a single gene, we might say something like 'I must have the gene that makes coriander taste awful'. However, 'gene' is the wrong word in this context. We have the same genes, but genes exist in different versions, and this is what creates our individual variation. A version of a gene is called an allele. In the example of the coriander-tasting gene, the main gene involved is called OR6A2 and it resides on chromosome 11. Its job is to make a protein called 6A2, which acts as an olfactory (scent) receptor, reacting in the presence of certain chemicals. A variant version of this gene produces a supersensitive version of 6A2, and people who have this variant allele on both of their copies of chromosome 11 find that coriander tastes strongly (and revoltingly) of soap. People with the 'normal' allele on one or both copies of chromosome 11 do not have this problem and can happily enjoy coriander-seasoned dishes.

Delicious or disgusting? When it comes to coriander, gene OR6A2 decides.

<< CHAPTER 4

The reason that this trait is present only in those with a double copy of the variant allele, and not those with just one copy, is down to dominant and recessive patterns of inheritance, which we discuss in detail in chapter 5. For now, just remember that when you have two different alleles at a gene locus and one is dominant over the other, you will only have the trait associated with the dominant allele – in this case the normal 'coriander tastes fine' allele. However, the recessive allele is still present in your genome and can be passed on to your offspring.

If your genetic make-up has one 'Coriander tastes good' allele and one 'Coriander tastes bad' allele, then coriander will taste good to you, because that allele is dominant and so is expressed, while the other, recessive allele is silenced.

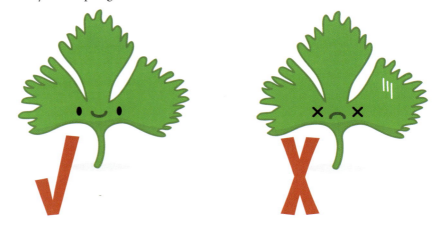

The same gene but different. Alleles are variants of a gene, and the two chromosomes of a homologous pair are likely to have a large number of unmatching alleles.

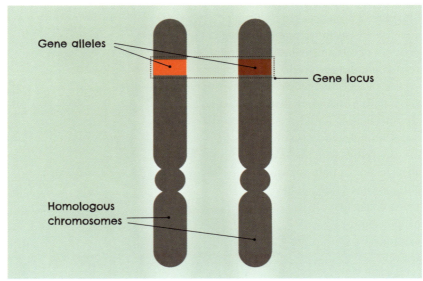

ALLELES >> 49

Many of our genes are known to exist as multiple alleles. An individual person can only ever have two different alleles per gene, as we only have two copies of each gene (one on each chromosome). However, across a population there may be more than two alleles in circulation. For example, the human blood group ABO system involves three different alleles of a gene that resides on chromosome 9. However, an individual person can't have all three – only two (or just one, if we have the same allele on both copies of chromosome 9). See chapter 5 for more about this gene. The HBB gene, which makes beta-globin (a component of the haemoglobin in our blood), is perhaps our most variable individual gene with hundreds of known alleles, although many of them have the same effect. We all have a unique combination of alleles in our genomes, and this is one of the reasons that we are all so different.

Because some cat fur colours are associated with recessive alleles, a litter of kittens may include some that don't 'match' their mother or their father.

MUTATION

New alleles arise through mutation – an error in the process of creating duplicate DNA. Although replication is a very reliable process, it is also a complex process and, with so many base pairs to copy, it's inevitable that the occasional copying mistake will happen. Each of us has, on average, about 70 new mutations that arose during the creation of the two gametes that formed us – something new that neither of our parents have in their genome. A mutation that results in a changed gene will mean that the protein that gene is coded to produce will be altered.

When we hear the word 'mutant' we might think of a superhero with a major and obvious difference to ordinary humans (usually, fortuitously, something that is incredibly useful in day-to-day life and crime-fighting). Sadly, in the real world, genetic mutations are much more likely to be either neutral or even harmful to the individual's survival chances, at least in the immediate term. However, if an individual organism does have a mutation that gives a distinct survival advantage in the environment where that individual lives, there is a fair chance that this new allele will spread through subsequent generations. We see this effect most clearly in lab studies of bacteria, as they complete their generational cycles so quickly. Bacteria in the 'wild' can also show rapid population-level changes in response to one or more genetic mutations, for example by developing resistance to antibiotics, or developing the ability to digest new types of molecules, such as nylon. These mutations are often passed on quickly through horizontal gene transfer, whereby one individual bacterium incorporates genes from another into its genome, through its cell membrane. (This process is also known to occur, albeit probably rarely, in eukaryotes.) Although mutations often have a neutral effect on survival initially, they might over time prove to be more useful, or more harmful, if the organism's environment changes. A mutation in a human population that results in an altered enzyme that is better than a previous version at digesting starch, for example, might make very little difference to those people's

Culturing bacteria in the lab under controlled conditions.

lives for generations, but would prove advantageous if that population's diet changed to become more starch-heavy (perhaps because they found a better way to grow a particularly starch-rich type of crop). A white-furred variant of a mammal that is usually brown would typically survive less well due to lack of camouflage – but it might fare much better if there was a sudden increase in snowfall (as long as it was also able to cope with the other challenges of living in a snowy environment). Famously, in 1975 Japanese biologists found a form of bacteria that was able to digest nylon, living in pools of waste water by a nylon factory – the bacteria produce three particular enzymes that allow this to happen, and each of the enzymes is coded by alleles that arose through simple errors that occurred during DNA replication. These new alleles could arise just as readily in any other population of that bacteria species, but would convey no benefits if there is no nylon around to be digested. However, as long as they do not harm the bacteria then they might persist in the population through the generations.

If a snowshoe hare's genes programme its fur to turn white when there is no snow on the ground, it is more likely to be spotted and caught by predators.

A change of direction

In the UK in the last decades of the 20th century, birdwatchers started to report seeing blackcaps in winter. These small warblers usually migrate to northern Africa or Iberia in autumn and don't return until the spring, so it was mystifying that they should switch their behaviour. Numbers of wintering blackcaps grew and grew, and thanks to ringing studies (where birds are marked with unique leg rings so they can be recognized again) ornithologists realized that these birds were not the same individuals as those that were breeding in the UK in spring and summer. They were actually breeding birds from eastern Europe, which would normally migrate south-west, to join UK breeding birds in Iberia and Africa, but were migrating west instead and ending up in the UK.

Lab studies on blackcaps from eastern Europe showed that the change of migratory direction was the result of a single genetic mutation. The new allele switched the preferred migration direction. The result, by a great stroke of luck, was that the blackcaps were ending up in a country with a relatively mild winter climate and a bountiful supply of food, courtesy of the 64 per cent of the UK human population who put out suet nuggets, seeds and other treats for garden birds in winter. Several decades on from the first reports of wintering UK blackcaps, the westerly migrators are showing a whole host of other differences from their south-westerly counterparts. Because the westerly birds have a shorter and easier migration back to eastern Europe, they arrive earlier and pair with each other, meaning that the two populations don't often interbreed, and other adaptations are appearing, as new mutations occur which are, in some cases, advantageous to the birds' new lifestyle (for example, browner colouration which helps them with camouflage in UK winter gardens).

TYPES OF MUTATION

There are several categories of error that can occur in the copying of DNA code, each resulting in a new allele. Deletion is when a part of the code goes astray, and insertion is when extra nucleotides are inserted into the code. Substitution is when a section of code is replaced by another of the same size. A frameshift occurs when the code is essentially corrupted because of an insertion or deletion that was not a multiple of three nucleobases, making all or most of the strand of code unreadable (because mRNA is read in 'words' that comprise specific codons of three nucleobases).

Sometimes a substitution has no effect at all, because each amino acid is coded by more than one different codon. A substitution mutation that changes the codon CCA to CCG will make a new allele, genetically distinct from the unmutated form or wild-type, but will have no effect on the protein produced, as both codons code for the amino acid proline. However, a new protein that differs in just one amino acid from its wild-type could be very different in function. The protein molecule haemoglobin, key to transporting oxygen in red blood cells, comprises 141 amino acids. In sickle cell disease, the sixth amino acid in the chain is substituted – glutamic acid for valine. The result is serious life-long illness, as the altered version of haemoglobin causes the shape of the blood cells to deform.

Types of mutations.

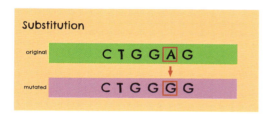

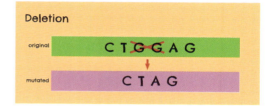

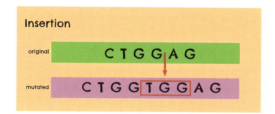

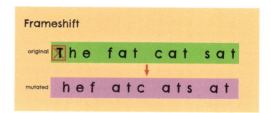

X-INACTIVATION

The X chromosome is a special case in terms of how its alleles are expressed. One familiar example is the primary coat colour gene in domestic cats, which has two alleles producing either black or orange (also known as red or ginger) fur. Have you ever wondered why all (or almost all) of the tortoiseshell and calico cats you see are female, and why their fur contains a mix of black and orange in equal proportions but always in a unique and asymmetrical pattern? The answer lies in the X chromosome, of which female cats, like other female mammals, have two copies. The gene determining primary coat colour resides on this chromosome and has two alleles – black (B) and orange (O). So females have one of three genotypes – BB, BO or OO. BB gives a black phenotype and OO an orange phenotype, but BO gives the random patchwork pattern of both colours that we call tortoiseshell (or calico if the cat also has white patches, which is determined by a different gene altogether).

Causes of mutation

Copying errors happen naturally by chance, in gametes and in somatic cells alike, but certain factors can make them more likely. Outside factors that can raise the rate of mutation are known to be *mutagens* – tending to promote higher rates of genetic mutation. Many mutagens are carcinogenic – tending to produce changes that lead to cancer, through the excessive proliferation of cells with the mutated genes. Exposure to radioactivity is a well-known example, and cigarette smoke, snuff and radon gas are others. Chemical mutagens damage DNA through chemical reactions with elements of the DNA molecule, while physical mutagens physically disrupt the structure of DNA. The effects of chemical mutagens may be somewhat mitigated by antioxidants, which can deactivate them. Other factors that can raise the rate of mutations include age, with the offspring of older parents being more likely to be affected by genetic problems.

Categories of mutation-causing agents (mutagens).

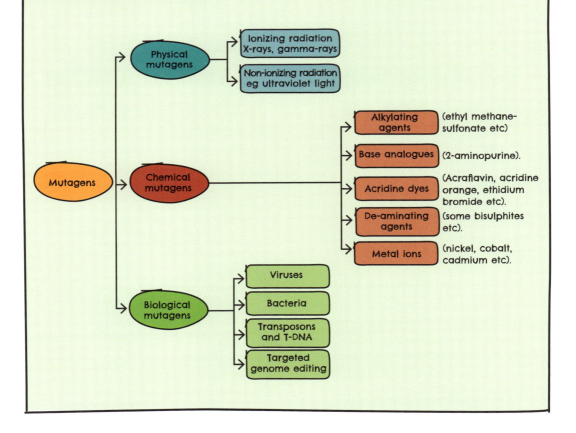

ALLELES >>

The random nature of the pattern is due to X-inactivation. In each cell of a female mammal, one X chromosome is randomly almost fully deactivated, being transformed into a small condensed mass called a Barr body. In the case of tortoiseshell cats, this means that some of the pigment-bearing skin cells express B, producing black hair, and others express O, producing orange hair. Genetically normal male cats, with only one X, cannot inherit both alleles and so can only be black or orange, never tortoiseshell. Because X-inactivation happens during embryonic development, it would mean that even identical twin tortoiseshell cats would have different coat patterns to each other. Because of X-inactivation, it is also possible for some females who are heterozygous for sex-linked diseases, such as haemophilia, to show some symptoms of the disease. See more about X-inactivation and Barr bodies in Chapter 8.

Note: *The Barr body is named after Murray Barr, who discovered the structure in 1948 together with his graduate student George Bertram, at the University of Western Ontario. X-inactivation is also known as lyonization, after Mary Lyon, who discovered the process in 1961 while working on radioactivity at Harwell, Oxford.*

X-inactivation also occurs in males who have the trisomy condition known as Klinefelter's syndrome, giving an XXY karyotype. This means that male cats with XXY chromosomes can be tortoiseshells. However, some reported male tortoiseshells have a condition called chimerism instead (see note below).

CHIMERAS

An organism that is a chimera has populations of cells that have two (or more) different genotypes in its body, and this has several possible causes. Just as an embryo can divide into two after fertilization, producing monozygotic twins, so two early-stage embryos can fuse into one, resulting in chimerism. This embryo will have two distinct cell genotypes as it develops, though this usually causes no obvious issues. One case of obvious chimerism is the phenomenon of 'half-sider' or bilateral gynandromorph, best known in birds and butterflies. This is the result of fusion between a male and a female

Many cat-lovers know that virtually all tortoiseshell and calico cats are female and most ginger/orange cats are male: the X chromosome holds the explanation for this.

embryo, in species that show sexual dimorphism (i.e. the males and females are clearly different in appearance). In these individuals, the male side and female side are divided down the centre line, which can be very striking in the case of butterflies which have a distinct size and colour difference between the sexes. Bilateral gynandromorphs are extremely rare, and cannot normally breed (even if they do produce gametes though, they could not pass on their unusual genetics as their individual cells are normal).

Note: *Chimerism can also result in a male tortoiseshell cat with a normal XY karyotype, if it originated from one male embryo carrying the B allele and another male embryo carrying the O allele.*

Because of X-inactivation, female mammals have chimerism naturally – some cells express one X chromosome and some the other. This type of chimerism,

Coats of many colours

Just as every cat carries alleles for either black or orange fur or both, every cat also carries an allele for one of several possible variations on the tabby coat pattern (for example, striped, swirly or spotted). However, an array of other colour-modifying genes are also at work in the feline genome, and because of a long history of developing and combining colour alleles through selective breeding, many domestic cats (and other domestic animals) express multiple colour-altering alleles at the same time. The tabby pattern, whichever it is, can be suppressed completely by an allele called nonagouti, which makes the individual hairs solid-coloured rather than having pale bands, as seen on the paler, rather shimmery-coloured fur of tabby cats in between their solid-coloured stripes. The white-spotting allele causes depigmentation over up to 50 per cent of the cat's body if the cat has one copy of it, and up to 100 per cent if it has two copies. Various other alleles can dilute the coat colours in different ways, by changing how pigment is deposited in the hairs. The allele causes a variety of albinism which is temperature-sensitive, resulting in depigmentation over most of the body, but with the cooler extremities (face, ears, legs and tail) remaining more fully pigmented. See chapter 7 for more on how we have used selective breeding to develop and combined varied alleles to create distinctive forms of domestic animals and plants.

Himalayan rat. This allele also reduces pigmentation of the iris, resulting in pink rather than dark brown eyes in the domestic rat.

where genetic changes happen within cells after fertilization and spread through the body as the cells proliferate, without any embryonic fusion occurring, is known as mosaicism. Mosaicism may also occur if one of the first cells (or blastomeres) mutates, and its descendant cells (which may then differentiate into many different types, proliferate all through the body) carry the new genotype, alongside the rest of the cells in the body which have the original genotype that formed at conception. This can only happen, however, if the mutation in question allows the cell to continue functioning normally – many such somatic mutations will render the cell non-functional and it will leave no descendant cells. This could also be fatal for the embryo itself, depending on developmental stage, although accidental (rather than pre-programmed – apoptosis) cell death in embryogenesis can occur without any issues for the embryo as a whole as other cells take over. Two clearly different eye colours in a human (heterochromia) can be the result of mosaicism.

Heterochromia.

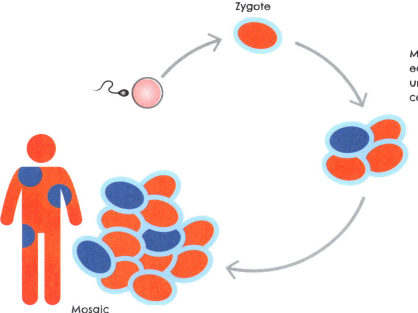

Zygote

Mutation in an early-stage, undifferentiated cell

A mutation at an early stage of embryonic development can result in an adult individual with populations of cells that have different genotypes: a genetic mosaic.

Mosaic

Chapter 5
Inheritance

- Dominance
- Sex-linkage
- Punnet squares
- Phenotype and genotype
- Prediction

YOU'VE GOT YOUR MOTHER'S EYES but your father's nose – or it could be the other way around. We know that we inherit traits from our parents, and that it's the same for other species too, but it seems very difficult to predict which traits we'll inherit and which we won't. The same goes for animals – when two breeds of dog are crossed, the puppies have a mixture of their parents' traits but can still look so different that you wouldn't necessarily guess they were related, let alone full siblings. However, where a single gene with a small number of possible alleles is concerned, inheritance patterns are very clear and satisfyingly predictable.

THE TALL AND THE SHORT OF IT

You have probably heard mention of 'dominant and recessive genes'. Once again, the right word would be alleles, not genes, because the terminology of dominant and recessive in genetics concerns the relationship between two (or more) alleles of a particular gene. The discovery of this interaction was first made by Gregor Mendel, an Austrian biologist and monk who carried out hybridization experiments on different varieties of domestic pea plants. He noticed that when he cross-bred tall plants with short plants, or tall plants with other tall plants, the resultant offspring were either tall or short, rather than a range of heights. However, crossing two short plants together produced only short offspring. He had discovered a trait that was controlled by just one gene, and that one gene had two different alleles (within this population at least) – tall and short, with tall being dominant over short. Because this gene resides on an autosome rather than allosome, the tall trait and allele are known as autosomal dominant, and the short trait and allele are autosomal recessive.

Gregor Mendel (1822–84).

Although Mendel's studies took place in the mid 19th century, before anything was known about genes and chromosomes themselves, his findings translate perfectly into what we know today.

He elucidated how the two alleles worked by first producing 'true-breeding' lines of tall plants (which only produced tall offspring), and then crossing them with the already true-breeding short plants and taking note of what happened in subsequent generations. He discovered that the first-generation offspring were 100 per cent tall, but when crossing these first-generation plants, the second generation were 75 per cent tall and 25 per cent short.

The first-generation plants each inherited a tall allele (which we'll call T) from the true-breeding tall parent and a short allele (which we'll call s) from the true-breeding short parent, so their alleles were T and s. With this combination, the dominant tall allele is expressed and the recessive short allele is not. Things worked differently in the second generation, though. Because each parent plant had one short and one tall allele to pass on, the offspring could inherit these alleles in one of four possible combinations. They could inherit T and T, T and s, s and T, or s and s. The offspring with the combinations TT, Ts and sT (which for our purposes is the same as Ts – it doesn't matter which parent contributed which allele) were all tall, because of allele T being expressed over allele s. However, one in four (25 per cent) inherited the combination ss, and were short, because there was no T present to be expressed.

GENOTYPE AND PHENOTYPE

Inheritance of alleles like these is best shown through a type of diagram called a **Punnett square**. For the case of the tall and short pea plants, Mendel's first-generation cross looked like this:

	T	T
s	Ts	Ts
s	Ts	Ts

The second-generation cross looked like this:

	T	s
T	TT	Ts
s	Ts	ss

INHERITANCE >> 61

So, TT and Ts are both tall – physically they are the same. However, they have different genetics. We express this by saying that they have the same phenotype, but a different genotype. We can't tell by looking at a tall pea plant whether it has a TT or Ts genotype, but we can tell if we breed from it. If we pair a TT with a short plant (which we know will always have the genotype ss), we know that all the offspring will have a T phenotype, and a Ts genotype. However, if it turns out that our tall plant is Ts rather than TT, the results of crossing it with a short plant will be different.

	T	s
s	Ts	ss
s	Ts	ss

In this combination, 50 per cent of the offspring will be tall and 50 per cent short (and we will know that all of the tall offspring have a Ts genotype, because they could only inherit one T, from their tall parent). We call the TT genotype homozygous in respect of the T/s trait, as both of its alleles are the same, while the Ts genotype is heterozygous. Another term you may see to describe a heterozygous individual with a hidden recessive trait is 'split' – this term is often used in animal breeding.

Nearly all of the more obvious or interesting traits in humans are determined by lots of different genes. Unlike Mendel's pea plants, we don't come in 'tall' and 'short' forms with nothing in between – our height is on a spectrum, and numerous genes are involved (as well as outside factors such as nutrition). Some other traits, such as eye colour, were once thought to be under the control of one gene, with a brown-eyes allele which is dominant over a blue-eyes allele, but we now know better – there are no fewer (and possibly more) than 16 genes involved. Some other traits often claimed to be under the control of just one gene include the ability (or not) to roll your tongue, whether your earlobes are freely dangling or attached, and whether you detect a strange smell in your urine after eating asparagus – but all of these claims have been debunked too. One trait that is known to be controlled

Who was Punnett?

You might be picturing a punnet full of freshly picked peas, but the Punnett behind the square is Reginald C Punnett, who devised this handy graphical way of showing outcomes of maternal and paternal allele combinations back in 1905. He was a British geneticist who, with his academic colleagues at Cambridge and St Andrew's Universities, helped to bring Mendel's work on inheritance into the scientific mainstream and to build upon his discoveries.

<< CHAPTER 5

by a single gene with one dominant and one recessive allele is earwax texture (wet or dry, with the wet allele dominant over the dry)!

Note: *You might have heard that, with eye colour, it is not genetically possible for two blue-eyed parents to have a brown-eyed child. In fact this is possible, as there are multiple modifying genes that determine eye colour, but it is very uncommon.*

CO-DOMINANCE AND INCOMPLETE DOMINANCE

With more than two alleles at a gene locus, the order of dominance is often linear, with one allele being dominant over the other two, and one of those remaining two being dominant over the third. This would be expressed as $B > b > b$ (with allele B dominant over b and b, and allele b dominant over b). In some cases, though, one allele at a gene locus is neither dominant nor recessive in relation to another. An organism with both alleles therefore fully expresses both traits and has a different phenotype again to individuals that are homozygous for either of the alleles. This is called co-dominance, and the ABO blood type system in humans provides an example. The ABO gene, which is located on chromosome 9, exists as three different alleles, which interact to produce six possible genotypes and four possible phenotypes. Many people will find out at some point in their lives what their blood type is, whether because they donate blood or need a record of it for possible future medical care, and a bonus of knowing this information about ourselves and our relatives is that it teaches us about a more complex inheritance pattern than the T/s pea plant example.

The alleles involved are known as A, B and O. O is recessive to both A and B, but A and B are co-dominant. This means that the heterozygous genotypes

		FATHER'S BLOOD TYPE				
		A	**B**	**AB**	**O**	
MOTHER'S BLOOD TYPE	**A**	A or O	A, B, AB, or O	A, B, or AB	A or O	**CHILD'S BLOOD TYPE**
	B	A, B, AB, or O	B or O	A, B, or AB	B or O	
	AB	A, B, or AB	A, B, or AB	A, B, or AB	A or B	
	O	A or O	B or O	A or B	O	

AO and BO result in blood types A and B respectively, and are phenotypically the same as the homozygous AA and BB genotypes respectively. A person who inherits an O from both parents will have the genotype OO and their phenotype will be blood type O. However, a person who inherits an A allele from one parent and a B allele from the other will have a fourth phenotype and blood type – AB. See chapter 9 for more about the implications of blood type genetics.

Incomplete dominance describes a case of one allele partially, but not completely, masking the effect of another, producing offspring that have an intermediate phenotype. The colour of ripe aubergines is controlled in this way, with a cross between a plant that bears white fruit and another that bears deep purple fruit resulting in offspring whose fruits are light purple in colour.

Knowing your parents' blood types helps you narrow down what yours might be, but you can only know for sure with 1 of the 16 possible combinations.

SEX-LINKED TRAITS

The X chromosome is one of the largest chromosomes in mammals, and carries genes that are related to a variety of functions. The Y chromosome, by comparison, is miniscule, with far fewer genes. This means that the genes on the X chromosome behave differently in females (with two copies of X) than they do in males (who have one X and one Y). The same goes for birds, although as we saw before, in their case it is males that are the homogametic sex (with two sex chromosomes that are the same, in their case ZZ) and females that are heterogametic (ZW). Traits governed by genes on the sex chromosomes are known as sex-linked.

Several of the colour variants that we see in domestic pet birds are sex-linked traits. One example is the colour 'fawn' in the zebra finch. The usual allele for colour, seen in wild zebra finches (and often known as wild-type relative to other variants) gives mainly grey plumage, while in fawn birds the grey is replaced by light brown. The gene concerned resides on the Z chromosome, and the allele for grey plumage is dominant over the allele for fawn. Therefore if a bird is heterozygous, with one allele for grey and one allele for fawn, its phenotype will be grey. To express the fawn phenotype, the bird must inherit the fawn allele on both of its Z chromosomes, one from each parent. This is only a possibility for a male, though, as only males have two Z chromosomes and so only males can be heterozygous and carry the fawn trait without expressing it. Females inherit one Z chromosome from their father and a W from their mother, so can only inherit a fawn allele from their father, and if they do they will always have a fawn phenotype. The W chromosome is irrelevant as it does not carry the gene concerned.

With Z representing the Z chromosome, W the W chromosome, G the dominant grey allele and f the recessive fawn allele, the Punnett squares for this type of inheritance look like this:

<< CHAPTER 5

Pair 1) Mother grey, father grey with a GG genotype

	Mother's alleles	
	ZG	W
ZG	ZGZG (grey phenotype male)	ZGW (grey phenotype female)
ZG	ZGZG (grey phenotype male)	ZGW (grey phenotype female)

Father's alleles

Pair 2) Mother grey, father grey with a Gf genotype

	Mother's alleles	
	ZG	W
ZG	ZGZG (grey phenotype male)	ZGW (grey phenotype female)
Zf	ZGZf (grey phenotype male)	ZfW (fawn phenotype female)

Father's alleles

Pair 3) Mother fawn, father grey with a GG genotype

	Mother's alleles	
	Zf	W
ZG	ZGZf (grey phenotype male)	ZGW (grey phenotype female)
ZG	ZGZf (grey phenotype male)	ZGW (grey phenotype female)

Father's alleles

Pair 4) Mother fawn, father grey with a Gf genotype

	Mother's alleles	
	Zf	W
ZG	ZfZG (grey phenotype male)	ZGW (grey phenotype female)
Zf	ZfZf (fawn phenotype male)	ZfW (fawn phenotype female)

Father's alleles

Pair 5) Mother fawn, father fawn with an ff genotype

	Mother's alleles	
	Zf	W
Zf	ZfZf (fawn phenotype male)	ZfW (fawn phenotype female)
Zf	ZfZf (fawn phenotype male)	ZfW (fawn phenotype female)

Father's alleles

INHERITANCE >> 65

As we can see, the only possible way to breed males with a fawn phenotype is from a fawn mother, and a father who is either homozygous or heterozygous for the fawn allele. These two combinations will also produce fawn females, but it is additionally possible to produce fawn females from a pairing of a grey mother and a grey (but heterozygous for fawn) father. This quirk of sex-linked inheritance explains why (all else being equal) phenotypes expressing sex-linked traits are more often seen in the heterogametic sex. That means there are more fawn female zebra finches than fawn male zebra finches. The blood-clotting disorder haemophilia in humans is another example of a sex-linked trait, controlled by a gene on the X chromosomes. In this case, because females are the homogametic sex, they are less likely to have haemophilia than males, but may carry the allele and pass it on to their sons (but not their daughters, unless they choose a partner who has haemophilia!)

'LETHAL GENES'

A variant that you may observe in domestic mallards is known as 'crested' and affected birds sport what looks like a feathery pompom on the backs of their heads. Charming though the crested form is, it has a dark secret. This trait is autosomal dominant, meaning that a duck that has one copy of the wild-type/ crestless allele (which we'll call c) and one copy of the crested allele (which we'll call C) will have a crest. If we pair two such ducks together, the Punnett square we would expect would look like this:

Mother's alleles

		C	c
Father's alleles	C	CC	Cc
	c	Cc	cc

In terms of phenotype, that would be 75 per cent crested offspring and 25 per cent uncrested. However, breeders soon discovered that the proportions were not as expected – instead, these pairings produced 66.6 per cent crested and 33.3 per cent uncrested ducks. They also noticed that a higher-than-expected proportion of the eggs did not actually hatch. The reason for this is that the CC genotype does not survive. The Cc ducks have a skull malformation that causes the head feathers to grow abnormally, and in the CC ducks, this malformation is much more severe and is not survivable. There are several other known alleles in a range of species which the homozygous form will not survive, including yellow coat colour in mice, the Manx (tailless) variant in cats, and the golden-leaved (*aurea*) variant in snapdragons. Alleles that are lethal in both homozygous and heterozygous forms are very rare, for obvious reasons – such alleles would normally be eliminated from a population as none of those with the allele would ever be able to pass it on. Huntington's

Crested domestic mallard.

disease in humans is an exception – although it is a lethal condition, it progresses very slowly, usually causing symptoms only when the sufferer has reached their 30s or 40s. They therefore may already have had children by the time they become ill and discover they have the faulty allele.

SEX DETERMINATION

In nearly all mammals, the allosome (sex chromosome) inherited from the male parent determines the sex of an embryo. Those with an X and a Y chromosome develop a male phenotype, while inheriting two X chromosomes produces a female phenotype, in the vast majority of cases (see chapter 9 for exceptions). Sex is therefore determined at the point of conception, by whether the sperm that unites with the egg carries an X chromosome or a Y chromosome (the egg contributes an X, because the female parent only has X chromosomes). The action of a gene called SRY, which appears on the Y chromosome, sets the newly formed embryo along a male developmental pathway.

Because sperm carry the decisive allosome, it's often argued that a human father 'decides' the sex of his children. This may be an oversimplification; if X-sperm and Y-sperm have different traits that would favour one or the other, depending on the circumstances of conception. However, biologists have examined sperm morphology and swimming speed and have not yet discovered any meaningful differences between X-carrying and Y-carrying sperm. None of the various methods prospective parents may try to naturally 'sway' the odds of producing one sex or the other, from trying to conceive at a certain point in the menstrual cycle to eating a special diet, have been shown to be effective.

As we have seen, sex determination in birds works the other way around, with males being homogametic with ZZ chromosomes, and females heterogametic with ZW. In this case, it is the chromosome carried by the egg rather than the sperm which determines sex. Also, there is no SRY gene – instead, two genes on the W chromosome trigger embryos to develop along a female pathway, although their exact function is not yet known. The ZW sex determination is also seen in some reptiles and fish, and a variety of invertebrates. Another variant is the XO system, in which only a single type of sex chromosome exists (X). Individuals with two X chromosomes (XX) are female and those with one (XO) are male, producing sperm that may or may not include an X chromosome. A wide range of invertebrates use this sex determination system, as do a few fish and a handful of mammal species, while its inverse, ZO (in which males are ZZ, and females ZO) is known in some moth species.

Some (girls) like it hot (or cold)

Specifically, the temperature at which the eggs are incubated in the case of certain reptiles. A lower or higher temperature produces mostly female young, while a median temperature produces mostly males. Crocodiles do not have allosomes — however, some other species of reptiles in which sex is determined by temperature do. One example is the lizard *Pogona vitticeps*, which has the ZW sex determination system. However, the genetic instructions of its allosomes can be overridden by incubation temperature, with most of the genetically male ZZ embryos developing as female if eggs are incubated at high temperatures.

Sex determination by temperature in reptiles:
a) sea turtles;
b) tuatara;
c) crocodilians.

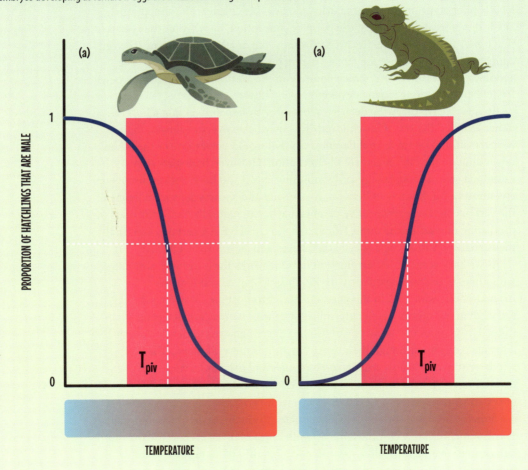

INHERITANCE >> 69

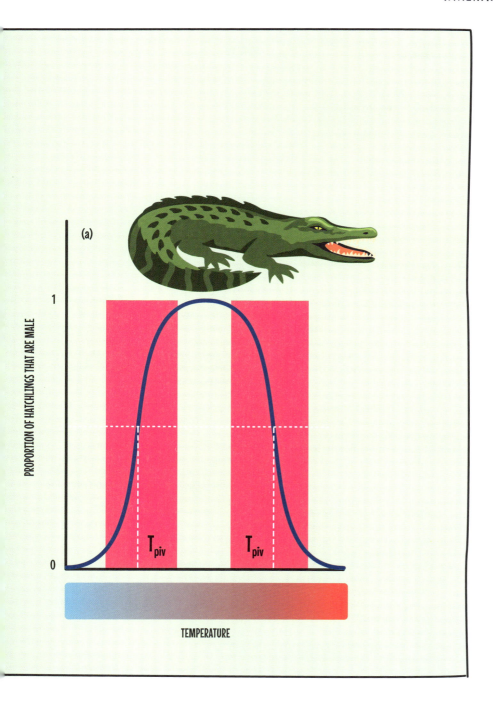

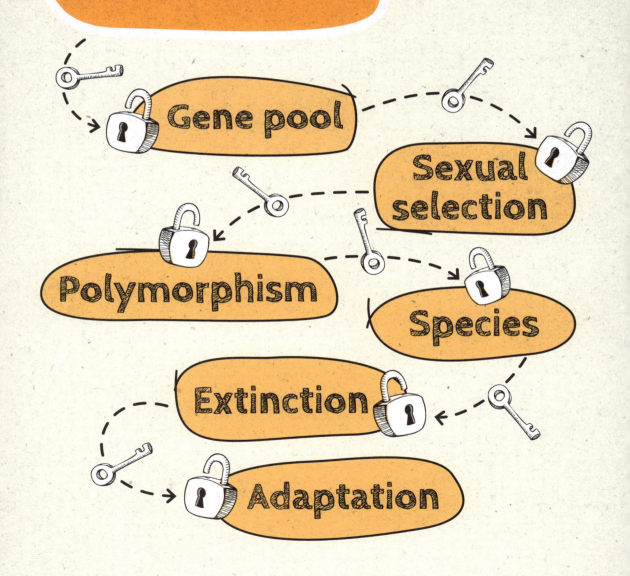

Many of us are familiar with the expression 'Survival of the fittest', which summarizes how the process of natural selection works. Only those individuals best adapted to their environment will survive, and only survivors will manage to breed. In this way, advantageous genetic traits are passed through the generations, and adaptation improves from generation to generation, with poorly adapted individuals dying, and insufficiently adapted species dying out. Environmental change may render useful traits suddenly less useful, though, and may also split one population into two, leading to the splitting of a lineage and the emergence of new species. Natural selection depends upon individuals having varied traits, and this is where genetics comes in.

GENE POOL PARTY

The term gene pool broadly refers to all the different alleles that exist within an interbreeding population. These generate the individual variation on which natural selection will act. The traits that are advantageous to survival are many and varied, and highly species-dependent. For a nocturnal, hunting animal, being able to quickly see changes in contrast is important, while for an animal that feeds on ripe fruit in daylight, colour vision is needed. For some animals, being naturally bold and adventurous in personality is useful, while extreme caution and vigilance is more useful for others. Some plants need to attract insect pollinators, so their flowers have an appearance and structure that appeals to and is navigable by insects. However, others have adapted to attract nectar-eating birds and bats, which requires very different flower shapes, colours and scents. All of these traits are under genetic control.

We can see tremendous variation in our own species, but to us, one impala, cornflower or *E. coli* bacterium looks much like another.

Different species of hummingbirds have bills that are specially adapted to feed from different shapes and sizes of flowers.

A mother scorpion is adapted not just to make babies but to nurture and defend them.

If the differences are as subtle as they appear to us, does natural selection really have that much variation to work with? We have to accept our human-centric bias here. All vulturine guinea fowl look the same to us, but the birds themselves can easily differentiate dozens of different individuals (and choose not only mating partners but preferred social companions on this basis). However, it is also the case that natural selection does tend to prune away more dramatic variation, shaping a species through its generations into the most survival-optimal form and creating a certain level of uniformity.

SEXUAL SELECTION

Surviving is only one part of the picture when it comes to evolution. An individual might have the most optimal genes possible to live and thrive, but those genes will die with their bearer unless it also manages to reproduce,

and its offspring inherit its well-adapted qualities. For the simplest, single-celled prokaryotes, reproduction is a straightforward process – the chromosome duplicates itself, and the cell divides into two, one containing each copy of the chromosome. For a complex organism that reproduces sexually, chromosomes must also be duplicated in order for reproduction to occur, but there is an array of other processes around this, both physical and behavioural. They may include (depending on the species) successful maturation of sexual organs, attracting an equally healthy breeding partner (while defeating rivals for that opportunity), going through a pregnancy, making a home for the young to develop, devoting time and resources to parental care, and perhaps even sacrificing one's own life entirely to help give offspring the best possible start in their lives. Just as with survival-enhancing traits, these anatomical, physiological and behavioural traits that support successful breeding are all under a selective pressure, and only those individuals that hold the 'best' alleles to facilitate reproduction will successfully pass on those alleles to a new generation.

If one parent can manage the parental role alone, then it does not need to choose a mate on the basis of good parenting behaviour – their health and vigour are more important. This leads to obvious sexual dimorphism, wherein the sexes look very different and members of the larger, more powerful or colourful sex (usually, but not always, males) engage in intense competition to demonstrate their fitness, often through physical confrontation with others of their sex. The other sex, meanwhile, needs to be fit in other ways, for other tasks. Think of technicoloured peacocks shaking their feathers to impress dowdy-coloured peahens, and bucks clashing

Male peacocks (below) and sage grouse (bottom) attract females with their fancy plumage. The females have camouflage as protection while they incubate their eggs.

antlers in between trying to keep control over a harem of does. The peahen needs camouflage while she incubates her eggs so being colourful would be a disadvantage to her, and the doe has no need to fight competitively as a strong buck can mate with many does, so she does not waste her precious bodily resources growing a new pair of antlers and massively strong neck muscles each year – instead her body is adapted to support a pregnancy and produce a sufficient milk supply for her fawn. In species that don't provide parental care at all, both sexes are free to select their partners on the basis of their apparent physical fitness alone.

Extreme examples of sexual selection can result in individuals with quite extreme physical or behavioural attributes that might impede their survival. The peacock may look splendid but his tail weighs him down when he takes flight to dodge an attacking tiger. The buck most strongly committed to fighting his rivals might fail to notice the stalking tiger at all. Devoting time and energy to performing well in courtship displays takes time away from other essential activities such as feeding and physical care. That sexual selection can be at odds with selection for survival places different types of allele in competition with each other, and the allele that prospers in one breeding season might be unhelpful in the next.

POLYMORPHISM

We often see uniformity in nature, but not always. In some species of damselflies, which are elegant predatory insects, the females occur in distinct colour forms – polymorphism. Male blue-tailed damselflies are black and blue, but females have three described colour morphs, the most frequent of which are *infuscans* (which is green), and andromorph, which is black and blue like the male. The fact that there are no females that are intermediate between these forms suggests the genetic mechanism behind them is simple, and so it has proved to be, with different alleles of a single gene determining which colour morph a female will be. The allele for andromorph is dominant

Battle of the sexes

Not all living things reproduce sexually, but for those that do require a partner in order to breed, the way they can best achieve the goal of passing on their genes as many times as possible may be at odds with the goals of their breeding partner. This conflict can set up a situation where the two sexes in a population will evolve increasingly elaborate behavioural and physical adaptations in response to each other. In many bird species, both sexes are needed for the hard work of successfully raising a brood of chicks, so they seek a committed partner. However, both sexes are also motivated to mate with other individuals outside of their established pair bond. The male benefits by fathering extra eggs in addition to those in his own nest, while the female benefits by introducing extra genetic diversity into her clutch of eggs. So both sexes are motivated to prevent their mates from straying, while also straying themselves. This leads to the evolution of behaviour like mate-guarding, females sneakily laying an extra egg or two in the nests of other females, and males ejecting eggs from nests when they have reason to suspect infidelity.

over *infuscans* and a third, much rarer pinkish form, *obsoleta*.

Although the andromorph is produced by a dominant allele, it is not necessarily the commonest form we see in the wild and in fact makes up just 4 per cent of all females in some areas, greatly outnumbered by *infuscans*. The explanation for this is that male damselflies will more quickly recognize *infuscans* as a female, so those of this form are most likely to mate. However, in years when the damselfly population is high, the *infuscans* females experience intense harassment from males which negatively affects their survival chances. In these years, andromorph females still manage to mate as there are abundant males around, but are more likely to survive the experience. Damselflies are more likely to occur in high numbers in warmer areas, so andromorphs are more frequent in these places, but both forms and therefore both alleles persist in the population across the species' whole geographic range, rather than one of them becoming 'fixed' in the population and the other all but disappearing.

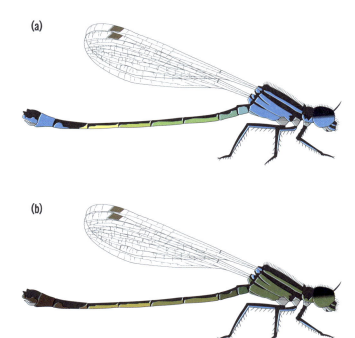

In the damselfly *Ischnura elegans* some females look just like males (a) while others are clearly different (b).

SPECIATION

As well as polymorphism in some species, we also see what's termed clinal variation. Particularly evident in species with an extensive geographical range, this form of variation refers to a gradation in a trait from one region or habitat type to another. For example, wolves living in colder climates are paler-coloured, thicker-furred and larger than those in warmer places, and the same plant species growing on a mountain will be shorter and slower-growing in the higher and more exposed areas.

As long as interbreeding occurs continuously across the species' range, clinal variation may not progress beyond this, but if a barrier to interbreeding

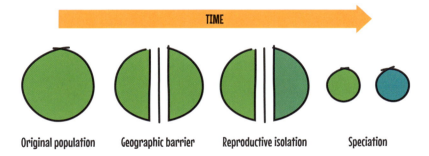

should develop (for example, climate change creating a new stretch of ocean that cuts a land-dwelling population into two) then speciation can occur. In this case, the two populations continue to breed and adapt through natural selection to their environments and, because the environments are different, there are differences in the 'best' alleles that become more numerous in the population over successive generations. Add enough generations and the two populations may become so different that they could no longer interbreed if reunited. This incompatibility could be at the genetic level, such as a change in chromosome numbers, but could also be introduced by a shift in behaviour or anatomy that means they no longer recognize each other as potential mates despite still being compatible in terms of chromosomes – this could be something as simple as a change from a two-note birdsong to a three-note version, brought about by a single genetic mutation.

INBREEDING

To return to our mental picture of a gene pool for a moment, you'll be aware that where this kind of pool is concerned, bigger is definitely better, as it would be if we were talking about swimming pools. A small gene pool means more chance that closely related individuals will breed together, and a lack of genetic variety, which in turn means that harmful alleles can be concentrated, and the resultant harmful traits are more likely to be expressed in individuals. It also means that the species as a whole is much more vulnerable to something like an outbreak of a new virus, as there is less chance of variants existing that are (by chance) naturally more resistant to the new threat.

With any very small population of animals or plants, inbreeding can become a problem. If a species is brought close to extinction, but then its numbers recover, it may still carry a legacy of issues from that population bottleneck. A well-known example in nature is the cheetah, which probably experienced two bottleneck events in its (relatively) recent past, one around

To split or to lump?

The Asian paradise flycatcher, a spectacular bird with a very long tail, can be red-brown in colour or pure white. The two forms are instantly, strikingly different to our eyes. Now consider two small brown butterflies found in Spain – Ripart's anomalous blue and Oberthur's anomalous blue. They appear almost identical, even if we dissect them, and the best way to be sure which one we have found is to look at a sample of its DNA. Yet the two flycatchers are the same species, and the two butterflies are different species.

How we decide whether two very similar living things are different species, different subspecies or merely different colour variants is based on a number of things – as well as physical distinctions we can look at whether they interbreed (and the results of that interbreeding), whether their geographical distribution overlaps or not, and of course we can now look at their DNA. However, even genetics can't tell us when to decide to split (treat as two species) or lump a population (treat it as one). We have to decide the cut-off for ourselves – we need to because so much of how we relate to, interact with and try to protect the natural world depends on a workable system of classification. Yet we also have to accept that it is, ultimately, unscientific and there is rarely a consensus, because evolution doesn't work that way. Single species do split into two, especially when an environmental change imposes severe selective pressure – an 'adapt or die' situation. Yet even then the change is gradual and incremental. Taking a cladistic approach goes some way to addressing the issue – see chapter 10 for more about this.

The arctic skua occurs in dark, light and intermediate colour forms, which all interbreed.

<< CHAPTER 6

Using coloured beads to represent alleles, this diagram illustrates how a population bottleneck can result in drastically reduced genetic diversity, even when population size has recovered.

100,000 years ago when a founder population reached Africa and Asia via land bridges (it travelled from the Americas – this archetypal African animal actually evolved on the other side of the world, and its closest relative is the puma). With a small founder population dispersing rapidly over a new area, these pioneering cheetahs had a sparse distribution, which limited their opportunities to exchange alleles.

Then, when the last Ice Age ended some 12,000 years ago, the cheetah population collapsed, with North American cheetahs disappearing completely and only a handful surviving in Africa and Asia. Numbers then rebounded considerably, but today the species has declined again for a range of environmental reasons, and the long history of inbreeding may be the final

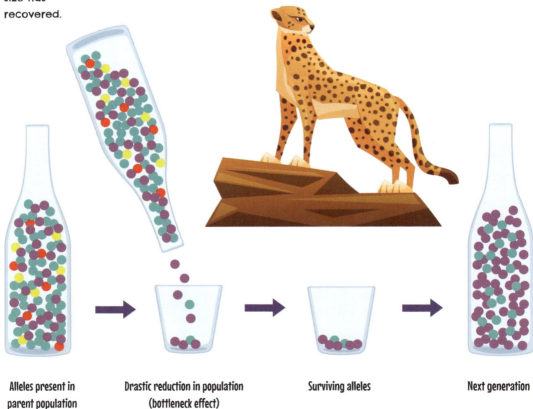

Alleles present in parent population   **Drastic reduction in population (bottleneck effect)**   **Surviving alleles**   **Next generation**

EVOLUTION >> 79

nail in its coffin. Cheetah genes reveal extremely high levels of inbreeding, with any two individuals having an average of about 95 per cent homozygosity (a measure of the amount of alleles they have in common, across their entire genomes). This is far in excess of other animals (for example, the average for domestic cats is 24.08 per cent, and even pedigree cat breeds with plenty of inbreeding in their history sit at around 60–65 per cent). The consequences of this lack of genetic diversity in cheetahs includes a high susceptibility to disease, and frequent occurrence of physical malformations, along with low breeding success in general. You might have read that any cheetah could probably accept a donated organ from any other cheetah because of their genetic similarity, which might be a small silver lining to the ominous inbreeding cloud (but it's unlikely to do much to save this troubled species from extinction in the long term!) A small gene pool will naturally grow with time as new mutations occur through the generations, some of which will be advantageous. However, this is a very slow process, particularly in larger animals which have a slow breeding cycle and don't produce many young at a time. In many cases the harms of inbreeding would cause extinction long before natural mutations could restore a healthier gene pool.

The human species has also survived population bottlenecks – most notably when small human populations migrated north from our species' birthplace in sub-Saharan Africa. Today, human populations in Africa still show markedly more genetic diversity than do non-Africans.

> ## Stud books
>
> Conservationists have saved many species from extinction through taking the miniscule remaining wild population into captivity and breeding them in safe conditions, with a view to returning some descendants to the wild in due course. Notable examples include the Chatham Island black robin, which was rescued from the brink thanks to the heroic efforts of just one male and one female robin, from which the entire current population of about 300 birds descends. Inbreeding is an inevitable consequence of such an operation, but can be minimized by carefully selecting breeding pairs. For example, the first generation of black robins would have had to be mated to full siblings or to parents, but in subsequent generations the conservationists can take care to pair only half-siblings rather than full siblings, then quarter-siblings and so on. A record of the family tree and the planning of pairings is kept in a stud book, and stud books are also used in developing domestic animal breeds, to concentrate the desired alleles while avoiding excessive inbreeding.

AN EVOLUTIONARY TIMELINE

We owe much to our picture of how life evolved on Earth to the field of palaeontology. Fossils show us what life-forms lived on our world and when, and by dating the rocks in which they are found, and comparing their structure to other fossils and to modern organisms, we can build a picture of evolutionary change over millions of years, from the first prokaryotes to the wonderful diversity we see today. Fossils are generally sparse and many are poorly preserved, so the picture is patchy. However, fossils can also give us clues beyond straightforward anatomical detail and distribution – it's sometimes possible to infer aspects of behaviour, such as locomotion, diet and even breeding habits from fossil finds.

Adding DNA to the picture allows us to refine that story much more, and place much more accurate dates against speciation events, particularly when it comes to working out when two modern lineages would have diverged. When we know the average mutation rate of a genome, this gives us a figurative

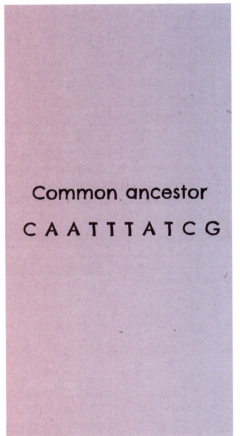
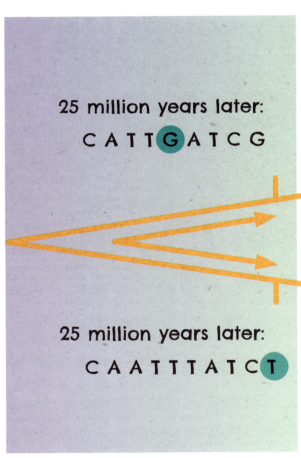

molecular clock for evolutionary change. We can then compare that genome with another in detail, and thus put a time-stamp on roughly how long ago it was that those two genomes were essentially the same. This is how we can be fairly sure that, somewhere in the region of 6.5–7.5 million years ago, the ancestors of chimps (genus *Pan*) and humans (genus *Homo*) were the same species. This hypothetical species, the most recent common ancestor or MRCA between chimps and humans, would have shared a more ancient MRCA with what became the gorilla lineage and, a bit further back, the orangutan lineage.

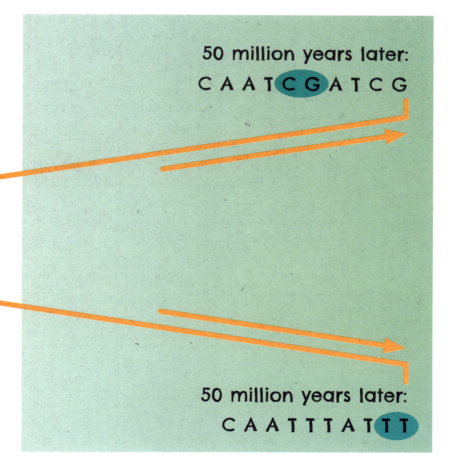

Based on known mutation rates, this graphic shows the degree to which a genetic sequence could change over time.

Chapter 7
Artificial selection and genetic engineering

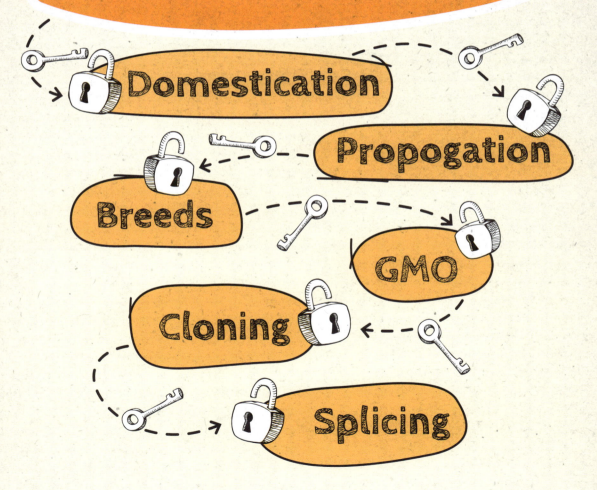

ARTIFICIAL SELECTION AND GENETIC ENGINEERING >> 83

EVOLUTION IS A NATURAL BIOLOGICAL PROCESS, with no plan or goal of any kind. It is also an inevitable consequence of what will happen when you have a population of living (and dying) things that self-replicate imperfectly. We humans have, historically, struggled to accept this truth, and it is perhaps not surprising that this is the case, given that we act with intentionality and that we have, over our relatively short history, become so adept at modifying nature to suit our own plans and goals. Modifying other species' genomes has long been important in our cultural evolution, since well before we even understood what a genome actually is.

SELECTION BOX

There are several species of wild dog in the genus Canis, the best known of which is the grey wolf (*Canis lupus*). Its fellow canids, which include the coyote and golden jackal, are all quite wolfish in appearance too, with pointed, upright ears, a long muzzle, rather a luxuriant tail, and coat colours that mostly combine various shades of grey, rufous and tawny-brown. There is one exception – *Canis familiaris*. This canid exists in a tremendously varied array of shapes, sizes and colours, and yet it is still sometimes classed as merely a subspecies of the grey wolf (*Canis lupus familiaris*) because the two are so closely related, with their most recent shared ancestor living just 14,000 or so years ago.

The domestic dog is a staggering example of what can be achieved through artificial selection. Given the timeframe over which we have turned wolves into pugs, greyhounds, poodles, spaniels, shar-peis and about 300 other distinct forms, the history of the dog and its genome dramatically highlights that, although evolution through mutation and natural selection is generally a very slow process, it does have the potential to bring about rapid change in the right circumstances. Artificial selection also reveals just how diverse and malleable a single species' genome can be.

Artificial selection works through selective breeding that focuses on particular traits. Most of the plants that we eat have a wild ancestor that is almost unrecognizable alongside its domestic variants. Once humans decided to cultivate crops, they could plant seed only from the plants that produced the biggest and more palatable fruit, and by controlling pollination they could breed the 'best' specimens together. Controlling matings between animals, so that we could choose which pairs we wanted to breed from, was

The genus *Canis* includes several closely related species of wild dogs. Domestic dogs descend from the grey wolf but, after many generations of selective breeding in captivity, domestic dogs are now genetically different enough from grey wolves that the two are usually considered to be separate species.

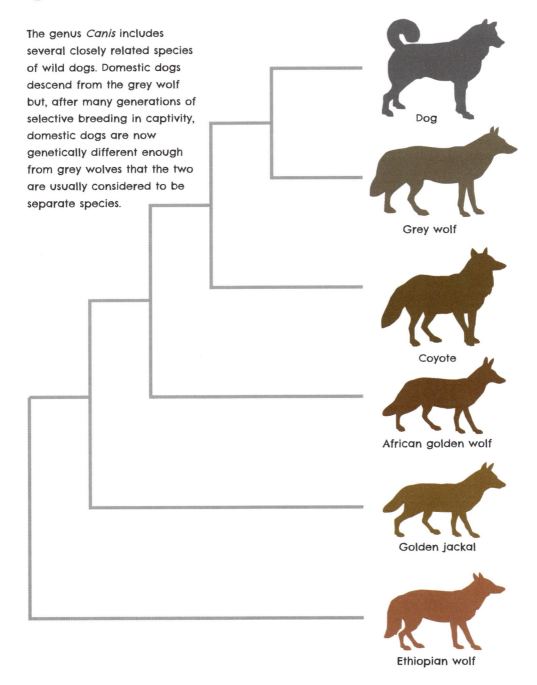

made straightforward once the animals could be tamed and contained. The process of domestication goes hand in hand with artificial selection, and it can take a surprisingly small number of generations for the desired traits to be considerably amplified. Breeders may also opt to create multiple lineages by focusing on different traits in different breeding groups, and their interest in creating diversity has given rise to the great variety of breeds, strains and variants that we see among the plants and animals that we keep today.

The relationship concerned may well be simply physical proximity on the genome. A mutation, by its nature, may well impact neighbouring genes (if nucleotides are inserted or deleted). Also, because recombination (see

A domestication experiment

When Dmitri Belyaev set out to domesticate a new species under controlled conditions, he selected silver foxes (a colour variant of the red fox) and chose his breeding pairs based on one trait alone – friendliness. The foxes were purchased from fur farms and were not used to human contact at all, but nevertheless showed variation in how they responded to people, and those that showed the least fear and aggression were chosen to produce the next generation. The experiment began in 1959 and, within just a few generations, had created a strain of foxes that were as cuddly as a pet cat. Something else was happening too, though. Even though no other traits were being selected for, the foxes were changing physically. Colour variants were emerging, such as white speckling or a 'tuxedo' pattern with white front and paws. The foxes' tails were becoming shorter and with a curl rather than straight, and their ear cartilage was softening, so the ear tips drooped rather than standing up stiffly. They also developed smaller brains, a longer-than-normal breeding season, and signs of neoteny (juvenile features being retained into adulthood). All of these are traits that have appeared in other domestic mammals, from cats and dogs to goats and rats. This experiment strongly hinted that a genetic basis for friendliness was linked to certain other variant alleles. In this case it seems likely that a gene locus associated with temperament has a relationship with the loci for other traits that are hallmarks of domestication – collectively known as *domestication syndrome*.

A tame domestic silver fox.

chapter 3) shuffles the genetic material but works in sections rather than necessarily cutting the chromosome into individual genes, it is likely that many neighbouring genes are still together after recombination. Therefore, genes involved in all of these traits probably do indeed reside close together on particular chromosomes.

CREATING DIVERSITY

Every so often, a breeder of animals or a grower of plants might discover a new oddity. It could be a kitten with stiff and curly fur rather than soft straight fur, a rose with a double flower, or a tropical fish with an extra-frilly tail fin. Almost every individual born in captivity with a brand new (de novo) genetic mutation and resultant new trait will face a very different future to that which would have faced it were it a wild animal. The curly-furred kitten would be less able to keep itself warm, the rose would be difficult for pollinators to access, and the fish would swim inefficiently and slowly – in all three cases, survival and breeding would be unlikely and the new allele would have disappeared. But in captivity, most oddities are prized and carefully line-bred to ensure that the new allele is passed on to multiple offspring. In this way, new forms of domestic species are created, and we enjoy the variety for its own sake.

Note: *When a breeder or farmer finds a new trait has emerged in their stock, and wants to propagate that trait, the first task is to work out the inheritance pattern of the new allele, relative to the original or wild-type allele. This means going back to Mendelian principles, and using line breeding (breeding from the new variant, and then breeding its offspring together, or back to the variant parent). Looking at the proportion of offspring from the first few generations that show the variant trait makes it possible to determine whether the new allele is recessive, dominant, partially dominant or co-dominant to other existing alleles, and whether it is sex-linked.*

Certain types of mutation have occurred in a wide range of domestic species. You have probably seen white pet cats, dogs, rabbits, mice, horses, ducks, geese, pigeons and budgerigars, to name but a few. These same domestic animals also come in variants with some white patches, and with a diluted version of their wild-type colouring. Genetic mutations that affect how melanin pigment (which is black, grey and brown in colour) is deposited in hair and feathers occur very frequently across many animal groups. In the case of birds that have carotenoid pigment (typically yellow) as well as melanin, a mutation that curtails melanin expression might have no effect on the carotenoid pigment, which is why many birds that are green in their wild-type phenotype may also be seen as an all-yellow phenotype in captivity – the canary is a familiar example. In the domestic cat, there are three different genetic pathways that can lead to cats that are entirely white, one of which is also associated with having blue eyes and, unfortunately, with deafness.

Note: *Most animals that are naturally not white in colour (i.e. express some pigment in their skin) have the potential to produce a white, partly white or paler-than-normal genetic variant, often in multiple ways in terms of the alleles involved. Darker-than-usual variants are also frequent in pigmented animals. However, darker variants of naturally white animals, such as most swan species, are almost non-existent. Intensifying, diluting or removing pigment that would naturally be there is a much simpler biochemical process than introducing it where there was none before.*

Most domestic mammals also have strains which have longer-than-usual hair, or curly hair, or no hair, and other very common traits include shortened limbs and face, and loss of tail. The genes involved are often the same between species – for example, the gene KRT71 has alleles that cause curly hair in cats, dogs, mice and rats (and humans, for that matter). By choosing their pairings carefully, breeders can produce animals that express a wide variety

A white blackbird and a white grey squirrel – both are lacking their species' normal pigment because of a genetic mutation.

of different non-wild-type alleles at the same time. For example, a wild-type domestic cat has a striped brown tabby coat of short hair, but a domestic cat could have alleles that turn the tabby pattern from striped to swirled, change the colour from brown to red (ginger), and throw in some white patches and make the fur long rather than short, all in the same individual. If a desired quality is controlled by multiple genes (for example, body size) then selective breeding can work to skew the range of alleles involved, such that successive generations are likely to be bigger (or smaller) than their parents, as required.

GENETIC ENGINEERING

Artificial selection has allowed us to manipulate the genomes of animals and plants, since long before we had any understanding of what a genome actually is. Now, though, we can take that manipulation to a whole new level, through genetic engineering, to create genetically modified organisms (GMO). Cloning is one of the best known kinds of genetic engineering. In many types of plants, it is possible to create clones with no special technology – you simply take a cutting. Animals that reproduce asexually create clones of themselves as a matter of course – no intervention needed. However, for more complex animals that reproduce sexually, the process requires multiple stages and a well-equipped lab. Dolly the sheep, the first mammal successfully cloned from an adult somatic cell, was created by removing the genetic material from a sheep ovum, replacing it with that from a somatic cell of a different sheep, and then implanting that ovum into a third sheep for gestation. Dolly therefore technically had three mothers – one gametic, one genetic and one gestational. This same technique has been used to clone many other mammal species. Birds, however, cannot yet be cloned, because they develop inside a hard-shelled egg – the ovum cell inside can't be taken out, manipulated and put back.

Note: *Dolly the sheep, who was born in 1996 at the Roslin Institute in Scotland, was named after the famous and famously busty country singer Dolly Parton, because she was cloned from a cell taken from a sheep's mammary gland.*

Editing actual genetic sequences is a more recent innovation. The first genetically modified plant that was approved for the commercial market was a variety of tomato called Flavr Savr, first sold in the US in 1994, and it included two additional genes. The first one inhibited the action of an enzyme that causes the fruit to decay, giving it a longer shelf-life, while the

Survival of the least fit?

Mutations that change an organism's appearance may have an impact on its health. The same goes for inbreeding, as we have already seen. Selective breeding, therefore, with its overriding goal of producing a lineage of individuals that look a particular way, can also create traits in that lineage that are inherently unhealthy, sometimes markedly so, relative to their wild ancestors. The world of domestic dog breeding is especially beset with controversy over the ethics around these issues. Health problems frequently associated with particular breeds include breathing difficulties in brachycephalic (short-faced) breeds such as British and French bulldogs, back pain in the long-bodied, short-legged dachshund, skin problems in breeds with loose skin folds, and a brain disorder known as syringomyelia in the Cavalier King Charles spaniel, because of the breed standard's desired head shape. While breeders work hard to eliminate genetically caused problems such as hip dysplasia from their animals' bloodlines, other more visible traits are considered inherent to a breed's identity. Breeders therefore face the challenge of retaining breed identity without compromising individual health, and many veterinarians would argue that certain breeds should be developed away from their present extreme forms. However, pet owners are often drawn to a particular breed precisely because of its highly distinctive (and, arguably, quite unnatural) appearance.

The French bulldog is a brachycephalic dog breed which can suffer breathing problems. Some other dog breeds are not able to mate naturally because of their body shape, and breeders must use artificial insemination to produce new puppies.

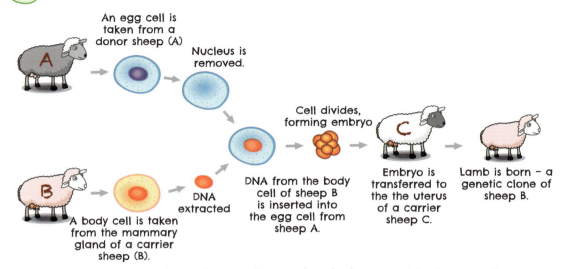

The creation of a cloned sheep.

second caused it to produce antifungal substances, giving it more resistance to crop-damaging infections. The genes were delivered via parasitic bacteria. The public were suspicious of this new tomato and it was a commercial flop. Genetically modified food products did not need to be labelled as 'GMO' under legislation at the time, but Flavr Savr was labelled as such. Despite this failure, subsequent GMO foods have fared better on the market, with public willingness to try them being bolstered by growing familiarity, lower prices, and reassuring laboratory testing with non-human animals.

Genetic engineering technology has a wide array of applications, not limited to those related to food production or medicine. It is even possible to genetically engineer organisms that can consume the most abundant and hazardous waste products that we generate through our other industrial activities. In chapter 10 we explore some of the newest technologies and look ahead to what may be possible in the future. The notion of genetic engineering is alarming in some ways, especially when we are not fully conversant with the facts (science fiction has a lot to answer for here) but carries the potential for facilitating dramatic improvements to our health, wellbeing, economic stability, and the long-term future of the planet we call home.

SPLICE AND DICE

We are able to read and interpret a genetic code with high accuracy, as we know exactly which codons produce which amino acids, and therefore which genes produce which proteins. We might also know the exact sequence of animo acids of a protein that we want to produce. So we will often know exactly what we want to change. The process of directly manipulating a section of genetic code could be compared to deploying the cut, copy and paste commands in word-processing software to change a sentence from one meaning to another, albeit working on a rather less user-friendly size scale (until we manage to develop intelligent nanobots to do it for us).

Splicing happens naturally as part of DNA transcription, as we saw in chapter 3. The purpose of this process is to remove non-coding sequences from newly generated messenger RNA, as well as to shuffle the coding portions in new configurations. The process is rather different in genetic engineering as it is DNA rather than RNA that is edited. The main way that gene editing is carried out involves the use of CRISPR (clustered regularly interspaced short palindromic repeats). These are repeated short sections of DNA that form parts of the genomes of many bacteria and function as part of the bacterial immune system, defending the bacterial cell against attack by viruses. When a bacterium is under viral attack, it adds pieces of the virus's DNA to its own genome, and these become CRISPR elements. They subsequently function to recognize the same virus again, and then make proteins that attack the viral DNA, cutting it into pieces (the process is often likened to attacking the DNA chain with a pair of scissors). CRISPR-based gene editing involves using CRISPR to snip the DNA in a cell at a specific point, and providing the cell with a new section of DNA that the cell will then use to repair the break.

Note: *Besides being a way to treat diseases, CRISPR-based gene editing helps us to determine the function of particular genes, by seeing what changes if a particular gene is tweaked.*

How CRISPR can be used to directly edit genes

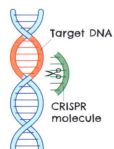

1 Search
A CRISPR molecule finds a precise location in the target DNA

2 Cut
The CRISPR enzyme cuts the target DNA at the point found by the guide

3 Edit
A new custom sequence can be added when the DNA is repaired

Chapter 8

Epigenetics

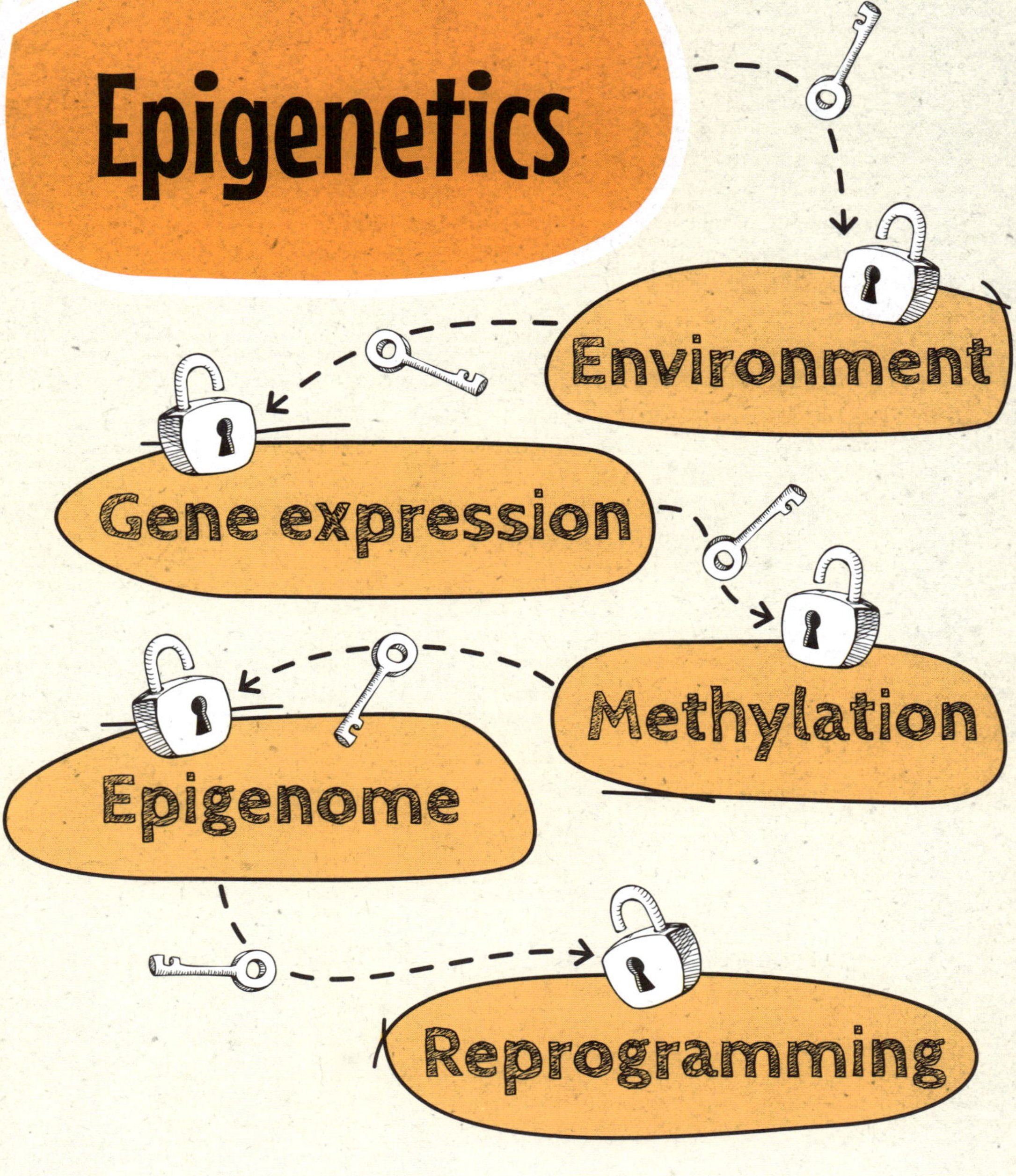

Environment

Gene expression

Methylation

Epigenome

Reprogramming

JUST AS WE CAN LEARN NEW SKILLS, we can change ourselves physically as we make our way through our lives. Put in enough hard workouts and your muscles will grow larger. Eat a healthy diet and you will have better digestive function and reduce your chances of developing various illnesses. Of course, you might also be careless with your health or your safety and cause damage to your body, and if you are deprived (whether of food, safety, education or love) in childhood your physical and mental state of health may be permanently affected. But it has long been observed and widely assumed that changes that are acquired during life are not passed on to subsequent generations – and this has been borne out by what we have learned about how genetic inheritance works. However, we are now discovering that things are not quite as simple as that.

NATURE AND NURTURE

The enduring debate about which of our traits are inborn and unchanging, and which are regulated by environmental factors, tends to conclude that 'it's a bit of both'. Environmental variations can make a huge difference to how an individual's life pans out, but the raw matter of who we are is dictated by our genes. Their code is set in stone and that code is the only thing we can be sure to pass on to our offspring, although we can do our best as parents to give them a healthy and opportunity-filled start to life.

This triggers questions, though. For example, we observe that adult height is influenced by the height of one's parents. However, we know from studies of identical twins who grew up separately that final adult height is not 100 per cent heritable. It can be as high as 80 per cent heritable, but the part of it that is influenced by environmental factors can add up to a lot – according

Identical or monozygotic twins share virtually all their DNA, so are great test subjects for 'nature versus nurture' studies.

<< CHAPTER 8

to the *Guinness Book of World Records*, the greatest known height variance between a pair of identical twins is 38 cm (15 in), after one of the twin girls suffered a developmental problem that caused dwarfism. Let us suppose that a man and a woman who are in a couple as adults had both suffered from severe malnutrition, or another non-genetic factor that seriously restricted their growth, during childhood or even before they were born. Would these hypothetical growth-restricted parents produce children that achieved the same potential heights that the parents' genes *should* have granted them, assuming those children had a problem-free upbringing?

Lamarckism

A contemporary of Charles Darwin (just about), Jean-Baptiste Pierre Antoine de Monet, chevalier de Lamarck (or just Lamarck from now on!) was a French naturalist who was interested in biological evolution, but also in natural history in general as well as medicine. He published many books and papers, in particular on invertebrate biology. However, he is most remembered (and often mocked) today for his views on evolution. Like Darwin, Lamarck believed that populations change through the generations according to natural laws, and he was the first person to develop a fully realized theory of evolution, published in his 1809 work *Philosophie zoologique*. However, his belief was that traits acquired in life could be inherited, and that animals were individually motivated to develop survival-enhancing traits. He called this drive 'adaptive force', and also described a 'complexifying force' which compelled organisms to acquire more complex forms and behaviours through the generations. Humanity represented the highest tier of progress (a view taken by many other naturalists of the time, to be fair).

Jean-Baptiste Lamarck (1744–1829).

Darwin's theory of evolution, published in *On the Origin of Species* in 1859, after Lamarck's death, put forward natural selection as the deciding force over which individuals lived or died and thus how adaptation and evolution occurred. After his ideas were widely accepted, Lamarck's contrasting view was often derided, as the real world failed to provide us with examples of children born with their parents' acquired traits (such as the muscular arms of blacksmiths). However, epigenetics shows us that acquired traits can indeed be inherited in some circumstances, and that the phenomenon is widespread and important in the evolutionary process.

SWITCHING ON AND OFF

Recent studies show us that the answer is no, not necessarily. Some early evidence of this came from a 2008 study examining the genes of people who were exposed to starvation while in utero. The cohort in question comprised people born in the Netherlands, during or just after the severe winter famine that affected the country in 1944–45. These people, even if they were a normal weight at birth, were more prone to health problems later in life. In particular they were prone to be overweight and to have associated problems such as higher rates of atherosclerosis (hardening of the arteries because of build-up of fats and cholesterol). These issues proved to be caused by changes in their gene expression, with certain genes being inactive or 'silenced' while others were active. The changes in gene expression also proved to be heritable, with the children and grandchildren of these individuals also having a tendency to become obese, while the progeny of same-sex full siblings who were conceived and born before and after the famine were not affected.

This phenomenon is termed epigenetics ('above genetics'). The term has been used under different definitions historically, but today it is usually used to describe all heritable traits, or changes to cell function, that occur in the absence of any changes to the organism's actual DNA sequence. We now know that epigenetic effects can affect any organism at any stage of life, and can be induced in lab studies (as well as observed in a variety of natural conditions). For example, researchers at Emory University, Atlanta, published a study in 2013 describing how male mice were trained with an aversion technique (a

Epigenetics is still a new science but we know that it is involved in many aspects of an organism's individual biology and life history.

mild electric shock applied to their feet) to fear the smell of cherry blossom. This brought about some observable and heritable changes to their DNA. Mouse pups fathered by the test subjects showed an immediate fear of the smell without experiencing any negative associations alongside their exposure to it, and the next generation also inherited the fear. Studies like this show how powerful epigenetic effects can be, and suggest a variety of (less unpleasant!) practical applications for induced epigenetic effects. The study of epigenetics and its causes and effects is still very young, but interesting discoveries are coming thick and fast.

HOW GENE EXPRESSION WORKS

Whether a gene is expressed or not in a particular cell comes down to a sort of chemical switch called a methyl group. This comprises one carbon and three hydrogen atoms (CH3), and it can be found as an addendum on various types of organic molecules. When a methyl group becomes associated with a gene on a section of DNA (a process called methylation), that gene can no longer make the protein it codes for – it is effectively switched off. Patterns of DNA methylation have been shown to be inherited through multiple generations – this includes induced methylation such as that observed in the sperm cells of the cherry blossom-fearing mice. The proportion of the genome that is methylated varies greatly between species, and also between cell types, with some cells typically very highly methylated and others hardly at all. Manipulating the methylation rate of certain cell types can have wide-reaching effects. The amount of methylation in the DNA within an individual mouse's adipocytes (fat-storing cells), for example, varies and is associated with body weight, and manipulation of the methylation process in these cells can induce or reverse obesity in that individual.

There are many possible causes for DNA methylation, besides those that are natural and part of normal development. As we have seen, starvation can trigger new methylation patterns, as can other physical stresses. Healthy dietary habits can cause other methylation patterns. Other environmental causes include variation in diet, and exposure to various chemicals. Because of the external nature of these triggers, the methylation patterns in identical twins begin to diverge at an early stage of development and will become increasingly dissimilar over time.

Another factor in epigenetics (and component of the epigenome) is histones – those proteins that DNA molecules wrap themselves around. Histones provide support, stability and protection for the DNA strand, but

The methylation process

In methylation, a methyl group attaches to one of the nucleotides that makes up a gene. In mammals, the vast majority attach to a cytosine nucleotide that is positioned just before a guanine nucleotide – this position is called a CpG site. An enzyme called DNA methyltransferase 1 (DNMT1) is involved in this process. *Demethylation*, the removal of methyl groups from DNA, is also facilitated by enzymes. Methylation is vital to deactivate unnecessary processes in embryonic cells as they begin to differentiate into distinct cell types with particular functions. While a stem cell has a high proportion of un-methylated cells, only 10–20 per cent of a fully differentiated cell's genes may be active, with different genes active and silenced in different cell types. Methylation can be seen as a guiding hand, ensuring a cell lineage follows the correct path and does not revert to a previous form. It is also important for deactivating retroviral DNA, which gets into the genome when a retrovirus's DNA is inserted and incorporated into a host cell's DNA. Demethylation restores activity to a gene.

Gene expression is affected by the addition (or removal) of methyl groups.

Cytosine

Methylated cytosine

also carry modifications that, like methyl groups, act as 'tags' to influence gene expression. Genes can be rendered physically unreadable by messenger RNA when they are held tightly wrapped around histones. RNA itself can also be methylated.

X-INACTIVATION

The deactivation of one X chromosome in each cell in genetically normal female mammals (with an XX karyotype), which we explore in chapter 4, may be cited as an example of epigenetics – a change in gene expression that happens without any changes to the actual DNA. However, it is not a heritable trait, as the mammal will still produce ova that bear an active copy of one or the other of her two X chromosomes, just as with her other chromosomes. In the somatic or body cells, the 'decision' as to which of the two X chromosomes is deactivated is considered to be random.

This is what was thought to be the case until recently, but we now know that patterns of X-inactivation can be heritable. There is also evidence that X-inactivation isn't as random as it appears, either, and it may be influenced by genetics. A study at the Fels Institute for Cancer Research and Molecular

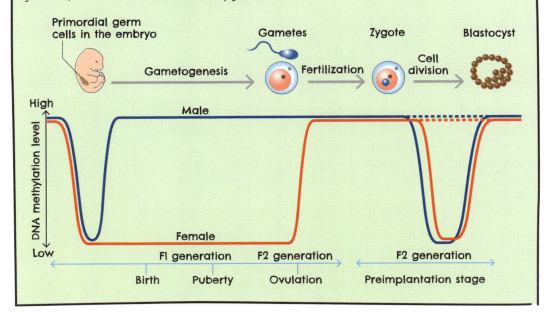

Clean-up

A newly formed zygote has very few epigenetic tags in its genome. This is important because this single cell is destined to divide many times, into populations of cells that will begin to differentiate into different types. Undifferentiated cells need to keep all of their genes active at this stage, but as differentiation progresses, un-needed genes can be silenced.

Sperm cells and egg cells are highly differentiated and specialized, and accordingly their genomes have numerous epigenetic tags. On conception, the tags need to be removed, and this occurs through a process called reprogramming. You will not be surprised, by now, to learn that this requires the action of specialized enzymes. Further reprogramming occurs at other points during embryogenesis, when certain cells need to be returned to their 'blank state'. However, about 1 per cent of the epigenetic tags in the genomes of the egg and sperm survive the reprogramming process. These are the source of heritable epigenetic effects.

Extent of methylation and demethylation of DNA during development in mice (over two generations).

Biology looked at families with many female members and found one example of a father whose seven daughters all shared a highly skewed pattern of X-inactivation, as did the man's mother. Potent genetic influence on X-inactivation has also been demonstrated in the lab, on mice.

GENETIC MEMORY OF TRAUMA

In human families, certain problems that are usually bracketed together as mental health issues, such as addiction and compulsive behaviour, often resurface in successive generations. It is easy to assume that this happens because of the social culture within the family, and that may well be a factor, but epigenetic changes are highly likely to play a significant role as well, making children born into the family predisposed to be much more vulnerable to these problems than their peers from other families.

Abuse can also bring about epigenetic change, for example by silencing the gene that encodes a protein (glucocorticoid receptor) involved in the stress response, leading to maladaptive behaviour when the individual is under stress. Use of psychoactive drugs can cause lasting epigenetic change in multiple genes. We should perhaps all pause to contemplate how easily these significant and heritable changes can come about. However, epigenetic change can also be undone, and this can be achieved through environmental means rather than physically altering the epigenome. Epigenetic research therefore offers the hope for development of effective new treatments for sufferers of a range of often devastating mental illnesses.

Barr body formation shown as part of X-inactivation, producing descendant cells with 50:50 mix of one or the other of the two original X chromosomes still active.

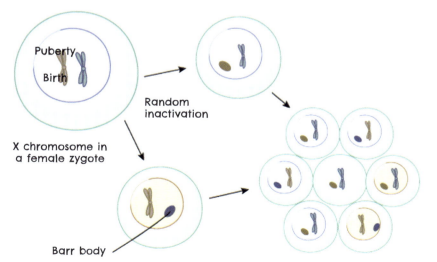

Mitosis of descendant cells after X-inactivation

Chapter 9

Genes and health

Monosomy and trisomy

Donation

Blood groups

Tissue-typing

P;Gene therapy

GENES AND HEALTH >>

FROM THE DISCOVERY that some health conditions run in families to tissue-typing and modern gene therapy, understanding genetics and inheritance has long been important in healthcare. Today, with the human genome fully mapped and genetic research forging ahead at high speed, people can attain a fuller understanding of their own genetic make-up and the strengths and susceptibilities that it grants them, and benefit from improved treatments for a wide range of conditions.

MONOSOMIES AND TRISOMIES OF AUTOSOMES

A frequent and serious problem that happens when gametes are formed during meiosis is for one of the newly formed chromosome pairs to fail to separate. This can happen at stage I or II of meiosis, with slightly different outcomes, but in both cases it results in gametes with one extra chromosome, or one missing chromosome. If a gamete with an extra chromosome then goes on to form a zygote, the zygote will have three copies of the chromosome in question – trisomy. If a gamete with a missing chromosome goes on to form a zygote, the zygote has just one copy of the chromosome in question – monosomy – instead of the two copies that it should have.

How trisomies and monosomies can arise during meiosis.

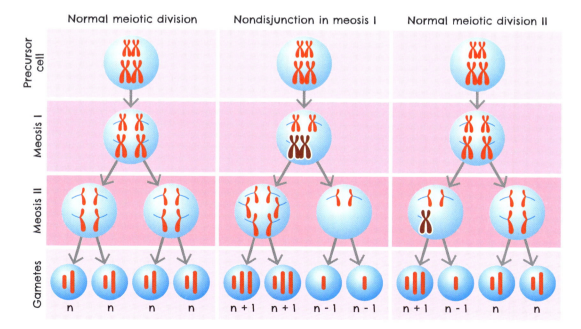

Monosomies of the autosomes (non-sex chromosomes) are lethal prior to birth. Trisomies of the autosomes are also usually lethal, but a few can be survivable. The best known is trisomy 21, which causes Down syndrome. Children with Down syndrome often have heart problems and significant intellectual impairment, as well as characteristic physical traits including short height, a flattened nose bridge and a distinctive eye appearance. However, the impact of the condition is extremely variable. Children with trisomy 13 (Patau syndrome) and trisomy 18 (Edwards syndrome) may survive to full term and live for a short time after birth but have very severe problems (although, again, prognosis is very variable).

Note: *It is possible to have a mosaic form of a monosomy or a trisomy condition, in which the individual's cells have a mix of diploid and monoploid/triploid chromosomes. This may result in less severe symptoms.*

MONOSOMIES AND TRISOMIES OF ALLOSOMES

Trisomies and monosomies of the allosomes or sex chromosomes have different outcomes to those affecting the autosomes. The monosomy XO (O indicating a missing chromosome) may result in a live birth. The child will develop along a female pathway externally, but without a functional reproductive system, and tend to have other problems including restricted height. The condition is known as Turner's syndrome. YO is not survivable – at least one X chromosome is always required. People with trisomies of the allosomes – XYY, XXY and XXX – often have few associated symptoms and have normal reproductive function, so they may never have reason to be investigated and diagnosed. XYY (Jacob's syndrome) and XXY (Klinefelter's syndrome) both develop along a male pathway, due to the SRY gene on their Y chromosome/s, while XXX develops as female. There are also rare cases of women with four X chromosomes – this condition, tetrasomy X, can produce a range of physical and mental problems. In women with more than two X chromosomes, all but one of the Xs will be inactivated leaving just one that is functional (just as in women with a normal XX karyotype).

Other rare problems with the sex chromosomes include a faulty, non-functional SRY gene in people with XY chromosomes, resulting in external development along the female pathway, but no functional gonads form. This is known as Swyer's syndrome. People with Swyer's syndrome do not experience puberty, as they have neither ovaries nor testes, and this condition also generally causes low bone density (osteopenia). Another very rare condition,

GENES AND HEALTH >>

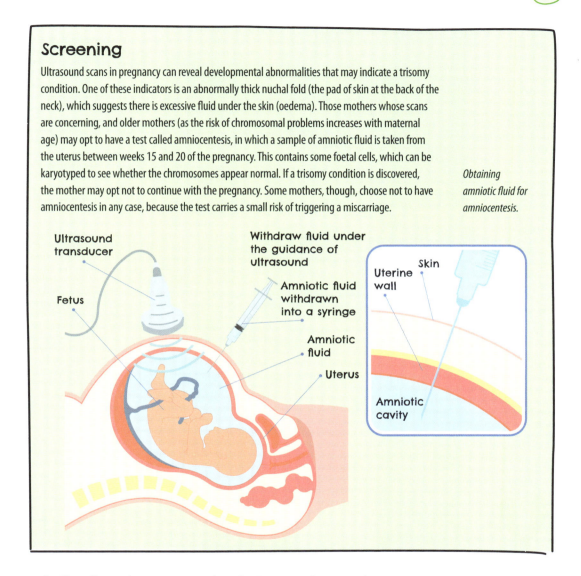

Screening

Ultrasound scans in pregnancy can reveal developmental abnormalities that may indicate a trisomy condition. One of these indicators is an abnormally thick nuchal fold (the pad of skin at the back of the neck), which suggests there is excessive fluid under the skin (oedema). Those mothers whose scans are concerning, and older mothers (as the risk of chromosomal problems increases with maternal age) may opt to have a test called amniocentesis, in which a sample of amniotic fluid is taken from the uterus between weeks 15 and 20 of the pregnancy. This contains some foetal cells, which can be karyotyped to see whether the chromosomes appear normal. If a trisomy condition is discovered, the mother may opt not to continue with the pregnancy. Some mothers, though, choose not to have amniocentesis in any case, because the test carries a small risk of triggering a miscarriage.

Obtaining amniotic fluid for amniocentesis.

De la Chapelle syndrome, occurs when the SRY gene from a Y chromosome becomes attached to an X chromosome during the formation of sperm cells, and that sperm cell goes on to fertilize a normal ovum. This usually results in the development of a functionally normal male phenotype, but karyotyping will reveal XX chromosomes.

<< CHAPTER 9

OTHER GENETIC DISORDERS

The genes known as BRCA1 and BRCA2 are strongly linked to breast, ovarian and some other cancers in humans. The former is found on chromosome 17 and the latter on chromosome 13, and mutations in either gene can greatly increase the risk of one of these cancers developing at a relatively young age. People with multiple relatives who have a diagnosis of one of the cancer types may choose (alongside their affected relatives) to have genetic testing to see whether they share one of the alleles known to be linked with cancer risk, and take preventative steps if they do. A woman who finds she has a dangerous allele may choose to have a double mastectomy and oophorectomy (removal of ovaries) to reduce her risk – she may also choose to have children before proceeding with the surgery.

For some genetic disorders, such as Huntington's Disease which is caused by a mutated allele on chromosome 4 that is dominant to the normal variant of the allele, there is no preventative care possible. Anyone with a parent who develops this condition has a 50 per cent chance of developing it themselves, though some of those at risk may prefer not to find out if they will develop the disease, as it is incurable. Prospective parents who know that they have a raised risk of passing on inherited conditions to children may opt for early amniocentesis, and those going though IVF treatment can have embryos screened for inherited conditions.

Examples of sex-linked genetic disorders in humans include Fragile X syndrome, haemophilia, and red-green colour blindness. Most sex-linked disorders only affect males and homozygous females, with heterozygous females being unaffected.

The genetic basis and physical progression of Huntington's disease, caused by a mutation in the gene HTT on chromosome 4.

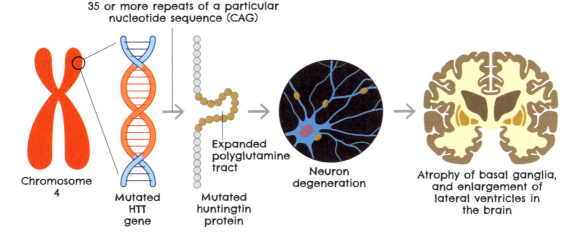

GENES AND HEALTH >> 105

DONATION OF ORGANS AND TISSUES

Catastrophic injury or disease can mean that the only way to save someone's life is by replacing a whole organ with one taken from a donor. Organ transplant surgery has existed since the 1950s, with the first successful kidney transplant in 1954 being followed by successful transplants of liver, heart and pancreas through the 1960s. By the end of 2022, more than 1 million organ transplants had been carried out in the US alone. Donations of this kind mostly depend on deceased donors. Next of kin may make this difficult decision, or the donor may have signed a register giving consent for donations after death, depending on the laws and protocols of their home nation. As people can live with one kidney and only part of their liver, these organs can be taken from living donors, as can donations of some types of tissue, such as bone marrow.

If the recipient's immune system recognizes the new organ as a foreign body, it will attack the organ's cells. This can be suppressed to some extent with drugs, but medical practitioners also carry out a test called HLA tissue-typing, to try to match an organ with the most genetically compatible recipient. We have many different antigens in the membranes of our cells. These are genetically coded proteins and are there to help our immune system recognize our own cells and distinguish any foreign cells (such as bacteria) they encounter. Six of these human leukocyte antigens (HLAs) are particularly important when it comes to donated organs and tissues. The genes that code for these six antigens have multiple alleles. Of the six, we inherit three alleles from each parent, so identical twins will always have the same six alleles and therefore make fully compatible versions of the antigens. However, full siblings may share anything from none to six. Having three or more shared HLA alleles is ideal, which is why close relatives are first choice for testing when a patient needs a living donor. In cases where a deceased donor is required, the tissue types of potential recipients are stored in a database and compared to the tissue type of each donated organ to find the most suitable match.

GENETICS OF BLOOD TYPES

People with blood type O have plasma antibodies against type A and type B blood, but no antigens on their red blood cells. Therefore they can only receive O blood, but can donate blood to those with any blood type. People with type A blood have anti-B antibodies and A antigens, so can receive type A blood as well as type O blood, and for type B the opposite is true – they have anti-A antibodies and B antigens, so can receive type B blood as well as type O blood. People with AB blood, having antigens for both A and B blood and no plasma antibodies of either kind, can receive blood from any donor but their blood can only be safely donated to others with type AB blood.

	GROUP A	GROUP B	GROUP AB	GROUP O
Red blood cell type	A	B	AB	O
Antibodies in plasma	Anti-B	Anti-A	None	Anti-A and Anti-B
Antigens in red blood cell	A antigen	B antigen	A and B antigen	None

GENES AND HEALTH >> 107

Blood cells may also bear Rh factor antigens on their surfaces (named after rhesus, as the protein was discovered through study of the blood of rhesus monkeys). Those who have the antigens are Rh positive, and those without are Rh negative. Rh status is determined by more than one gene, and there are numerous different antigens in the Rh system as a whole. However, the most important antigen in this system is known as D, and its presence is determined by the gene RHD on chromosome 1. A functional RHD allele gives Rh positive blood, and an absent or mutated variant gives Rh negative blood. Combining a person's ABO type with their Rh status gives eight possible phenotypes and 12 possible genotypes:

In terms of inheritance, Rh positive is dominant over Rh negative. The Rh factor affects blood donation alongside ABO blood types, as Rh negative blood makes antibodies in the presence of Rh positive blood. So Rh negative donors can only receive Rh negative blood (but can donate to both Rh

GENOTYPE	PHENOTYPE
AA +	A +
AO+	A+
BB +	B +
BO+	B+
AA–	A–
AO–	A–
BB–	B–
BO+	B–
AB+	AB+
AB–	AB–
OO +	O+
OO–	O–

Opposite: The antigens and antibodies present in the blood of the four human ABO blood types. Antibodies attack corresponding antigens.

Left: Human genotypes and phenotypes, by ABO blood type and rhesus status.

*Opposite:
Blood donor
and recipient
compatibility
chart – X
denotes
incompatibility.*

positive and Rh negative recipients).

A pregnant woman is often exposed to her baby's blood cells during pregnancy and while giving birth. If the mother has Rh negative blood and her baby has Rh positive blood, her blood cells will produce antibodies that attack Rh positive blood cells. This sensitization means that if her next pregnancy is also with an Rh positive baby, her antibodies can cross via the placenta and attack the new baby's blood cells throughout the pregnancy, causing rhesus disease (symptoms include newborn jaundice).

It is routine in many countries to determine the blood type of all pregnant women and to provide Rh negative women with injections of anti-D immunoglobulin, which will destroy any Rh positive blood cells that enter her bloodstream before antibodies can be created. An Rh negative woman who has already been sensitized to Rh positive blood will need careful monitoring throughout all of her pregnancies.

> ## Rarities
>
> It is usually not possible for a person with type O blood to have a child with type AB blood, because AB occurs when a child inherits an A allele from one parent and a B from the other, and everyone with type O blood only has O alleles to pass on to their children. However, there is a very rare variant of type AB, known as *Cis*-AB, in which both A and B antigens arise from a single allele, rather than from one A allele and one B allele. If a child inherits one *Cis*-AB allele from a parent, they will have type AB blood, but will also have a second ABO allele from their other parent which may be any of A, B and O (or even a second *Cis*-AB allele!) This variant, although exceedingly rare worldwide, occurs with more frequency in eastern Asia. Probably the rarest blood type is 'golden blood' or Rh null, which has none of the many different RH antigens at all. There are thought to be only about 50 people worldwide with this blood type, and they can only receive donated blood from others with Rh null blood.

BLOOD DONATION

The first successful blood transfusion took place in 1818, long before blood types were known about. Dr James Blundell, who carried out the procedure, also discovered that only human blood could be donated to human patients (which suggests that he was also responsible for a number of unsuccessful blood transfusions). The different main blood groups were discovered in 1901, and in the 1930s a method was found to store blood in a transfusable

GENES AND HEALTH >> 109

RECIPIENT	DONOR							
	O–	O+	A–	A+	B–	B+	AB–	AB+
O–	🩸	X	X	X	X	X	X	X
O+	🩸	🩸	X	X	X	X	X	X
A–	🩸	X	🩸	X	X	X	X	X
A+	🩸	🩸	🩸	🩸	X	X	X	X
B–	🩸	X	X	X	🩸	X	X	X
B+	🩸	🩸	X	X	🩸	🩸	X	X
AB–	🩸	X	🩸	X	🩸	X	🩸	X
AB+	🩸	🩸	🩸	🩸	🩸	🩸	🩸	🩸

condition, leading to the world's first blood banks being established in 1937.

In a medical context it is very important which blood type a person has, as the red blood cells of those with certain blood types will possess antigens that trigger immune attack against the antibodies present in 'alien' blood. This means recipients can only receive donated blood of the correct type, or they will become seriously ill through this immune response. The only blood type which can be accepted by any donor is O Rh negative. People with this type are especially valuable as donors, as a recipient can always receive O Rh negative blood even in an emergency where there has not been enough time to ascertain their own blood type. Once blood type has been identified, patients will always be given blood that matches their own, to save stocks of the precious, universally acceptable O Rh negative blood for emergencies.

Blood types around the world

The frequency of different blood types varies geographically, suggesting that there are evolutionary advantages to one over the others in different environments. It is known that blood type can affect vulnerability to different infectious diseases, and the idea of certain diets better suiting certain blood types has also become popular (although hard scientific evidence is still lacking here).

Worldwide, the proportions are:

O+	39.81 per cent
A+	27.40 per cent
B+	20.86 per cent
AB+	5.65 per cent
O-	2.68 per cent
A-	2.10 per cent
B-	1.14 per cent
AB-	0.37 per cent

O+ is particularly dominant in South America, with more than 70 per cent of Peruvians and Ecuadorians having this blood type. However, in south-eastern Europe A+ is most frequent, while B+ is well represented in the Indian subcontinent, and AB+ accounts for more than 11 per cent of people in North Korea (with A+ and B+ both outnumbering O+ there).

GENES AND HEALTH >> 111

The Austrian/American biologist Karl Landsteiner is known as the 'father of transfusion', for his work on developing the modern classification of human blood groups in 1901.

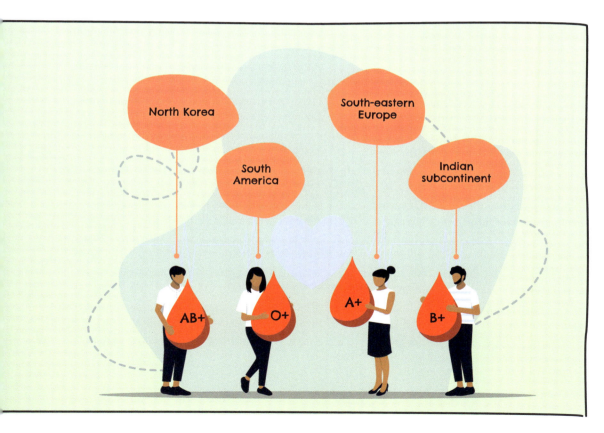

GENE THERAPY

For some inherited conditions, it is now possible to have treatment that involves actually altering the genetic code within cells, using gene editing techniques as described in chapter 7. In 2023, two new gene therapies were approved in the US by the Food and Drug Administration (FDA) to treat sickle cell disease. This is a genetic disease in which red blood cells are malformed, being crescent-shaped rather than flattened discs. The shape change is caused by a mutation in the gene that makes haemoglobin, the molecule that is involved in transporting oxygen to body tissues. People with the allele that causes sickle cell disease produce an altered version of haemoglobin A, and have severe problems with anaemia and associated symptoms.

One of these therapies, Lyfgenia, involves editing the genomes of the patient's stem blood cells (precursors to fully formed red blood cells) to produce a different variant of haemoglobin known as HbAT87Q, which functions like haemoglobin A but is less affected by the sickle cell mutation. The other therapy, Casgevy, works in a similar way but in this case the stem cells are edited to produce HbF or foetal haemoglobin, which is also resistant to the sickling effect. Gene therapy research is working on treatments for a range of other genetic conditions, including cystic fibrosis, but genetic treatments could also be developed for conditions that are not necessarily heritable, such as Parkinson's disease.

Magic bullets

The nickname 'magic bullet' for a treatment that is very highly targeted (for example to cells in a tumour, without harming the surrounding healthy cells) is usually applied to medications. However, some genetic therapies also carry the nickname. T-cells (along with B-cells, macrophages and others) are lymphocytes or white blood cells, and important parts of the human immune system, which reside in our blood and recognize and target specific threats. They do this through identifying certain proteins called antigens on foreign cells' surfaces, then binding to those cells and killing them. They also communicate with antibody-producing B-cells which also attack the foreign cells. Thanks to T-cells, if our system encounters a threat more than once, our immune system 'remembers' the first time, because a population of the right kind of T-cells that were active in the first infection are still around. They replicate, creating a T-cell army. The immune system is thus more effective at quickly neutralizing an infection on future exposures. It is possible to genetically modify T-cells to mount a strong attack against any cell bearing a particular antigen, including certain cancer cells, which may otherwise be ignored by the immune system as they are not recognized as dangerous. Genetically modified T-cells have been used successfully to tackle various forms of lymphomas – cancers that affect different cell types in the lymphatic system. This treatment is carried out primarily on patients who have suffered a relapse after conventional chemotherapy treatment. It involves collecting and culturing the patient's own T-cells, and modifying their DNA to 'teach' them to recognise the specific lymphoma cells the patient has, before transfusing the modified cells back into the patient's body. The treatment is currently risky, as it can trigger various serious side-effects including excessive production of the cytokine proteins that are involved in immune processes. A 'cytokine storm' can send the immune system into overdrive and cause systemic problems.

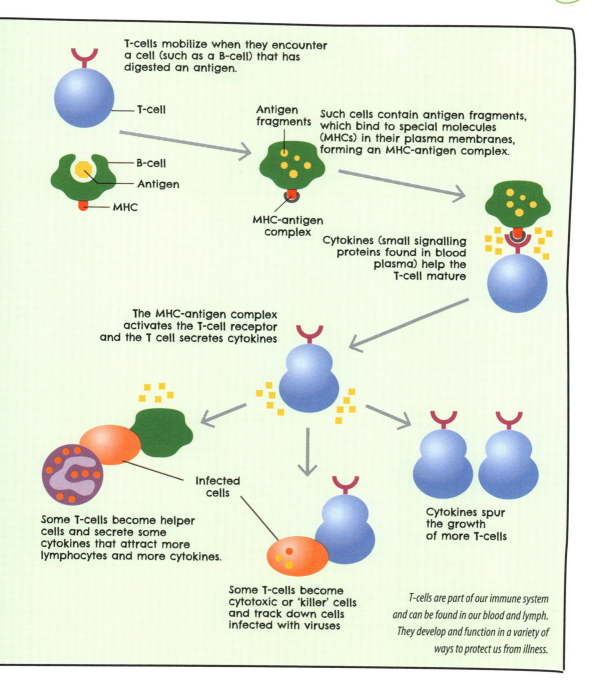

Chapter 10
The future of genetic science

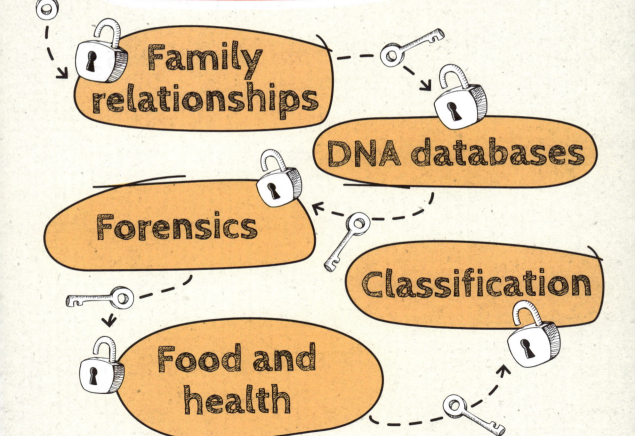

THE FUTURE OF GENETIC SCIENCE >>

GENETIC SCIENCE HAS COME A VERY LONG WAY in a very short time. Now, it impacts our lives in all manner of ways, from the medicines we take to the food we eat, and the colours of our garden flowers to the temperaments of our pets. Every branch of science that is concerned with life has multiple links to the science of DNA, genes and chromosomes, and these connections are only going to become more numerous and more important in the future.

GENEALOGY

DNA testing to determine paternity in humans has existed since the 1990s. Prior to that, there were cases of possible paternity being ruled out by comparing the blood types of the prospective father and the mother with that of the child, as some outcomes are impossible (for example, a couple who both have type O blood can only have children with type O blood, and a man with type O blood cannot be the father of a child with type AB blood, regardless of the mother's blood group). This is a form of genetic paternity testing, as blood type is determined by the genes, but in many cases it would not give an unambiguous answer. Modern tests, however, are extremely accurate.

Modern DNA testing to locate possible relatives not known to a person is now commonplace. It can also offer insights into a person's likely ethnic mixture, and their likelihood of developing certain illnesses. The tests involve looking at particular parts of the genome and comparing to databases and patterns established from large-scale analysis. Companies that offer this service commercially may also provide testing of mitochondrial DNA (which can be carried out on both sexes, and provides information about the female line of descent) and Y-chromosome DNA (which can be carried out on males,

Genetics can be used to trace your family tree, and also work out the combination of breeds in your pet dog's ancestry.

Digital DNA

Each year, more and more people submit a sample to one of the various DNA testing companies, in order to learn more about their personal genetic profile. As the majority of people using these services will opt in to having their data shared with research partners, the collective database of human genomes available to researchers is rapidly growing. Access to this wealth of individual genomic information is already having a very significant impact on many branches of academic and medical science. Professional genealogists, whose work involves tracing people who are, unknowingly, the closest kin to someone who has died without leaving a will, are among the winners in the commercial sector.

and reveals details of the male line of descent). DNA testing is also available for livestock and companion animals, to provide insights into the make-up of a mixed-breed dog or cat, for example. For rare animals or plants which are being bred in captivity to increase their populations, genetic testing can help determine the best pairings, to maximize genetic variation through the generations.

LAW

Our fingerprints are as unique as our genomes, and many a criminal conviction has been secured on the basis of fingerprint evidence. However, a careful thief or murderer could easily ensure they leave no prints on anything. Today, law enforcers and forensic scientists have another, much more powerful weapon in their armoury. DNA resides in all of our cells, and can be recovered from even miniscule traces of ourselves that we leave behind – even a hair, a flake of skin or a speck of saliva on a drinking glass could be enough to identify a perpetrator.

In England and Wales, anyone can offer to give a DNA sample to police. Many choose to do this, if invited, to eliminate themselves from enquiries around a particular crime where DNA evidence was recovered from the scene. Police also have the power to take and keep a sample from anyone who is arrested for any recordable offence (crimes for which the police must keep records – most of these have the potential to attract a custodial sentence). The countries' National DNA Database (NDNAD, also known as the National Criminal Intelligence DNA Database) now holds DNA samples from more than 3.5 million people, making it possible to identify a suspect even if no circumstantial evidence links them to the crime in question. It also opens up the possibility of re-investigating 'cold cases' that pre-date the inception of the NDNAD. There is much regional variation around how governments approach this issue. With prevention of all crime being the desired goal, and the potential of DNA detection acting as a significant deterrent, many citizens support the idea of their country's government keeping a DNA database, but the issue also raises concerns around personal privacy – it is difficult to imagine anything more personal than one's own genome!

THE FUTURE OF GENETIC SCIENCE >> 117

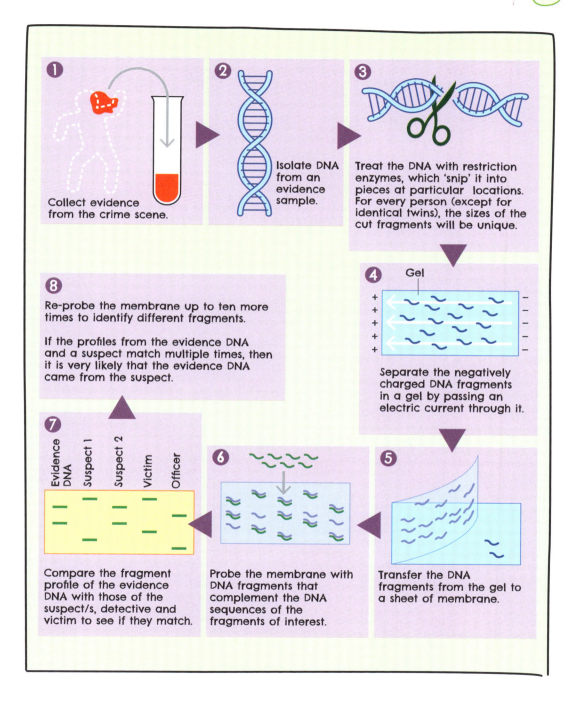

TAXONOMY

The study of taxonomy, or the classification of living things, might make you think of museum cabinets full of dusty specimens, but modern taxonomy is increasingly guided by the findings of genetic studies. Since the late 1990s, numerous long-established views about the arrangement of life's family tree have been overturned, and diagrams showing which groups of organisms descended from which have been redrawn (or discarded completely). For example, we might look at an owl, a hawk and a falcon and think they must be close cousins, and our old bird book agrees. But thanks to genetics we now know that while the owl and hawk are indeed close cousins, the falcon emerged from a very different lineage and is in fact a close cousin to the parrots.

Cladistics

Traditionally, living things are classified in a hierarchical series of levels. For our own species, that looks like this:

Domain: Eukarya (eukaryotic life, with complex cells containing a variety of organelles)
Kingdom: Animalia (animals)
Phylum: Chordata (animals with a spinal cord, most of which have a body skeleton and so are vertebrates)
Class: Mammalia (mammals – animals that have hair and internal thermoregulation, and produce milk for their young)
Order: Primates (the group of mammals that includes lemurs, monkeys and apes)
Family: Hominidae (the great apes, including chimps, gorillas, orangutans and humans)
Genus: Homo (humans, including extinct species such as *Homo neanderthalis*, Neanderthals)
Species: *Homo sapiens* (modern humans)

The problem here is that evolution is a slow, continuous process, and does not have any mechanism or effective indicator for separation of a population of organisms into a neat group. It's like looking at a tree – you can see the separate branches and twigs, but you cannot define the exact point where one branch or one twig divides into two. The solution of creating intermediate groupings (such as suborder or superfamily) only kicks the problem down the road. The solution is cladistics, which works on evolutionary units called clades. A clade contains one common ancestor plus all of its descendants. Thus *Homo sapiens* is a clade, containing the first ever modern humans and all their dead and living descendants. Primates is also a clade, containing the first true primates and every other animal that evolved from them, extant or extinct. But a clade could be any other subset within that group, provided it includes a whole genetic lineage. So, chimps and humans also form a clade, as their common ancestor existed after all other apes had already branched off, and all apes and Old World monkeys also form a clade, because their common ancestor existed after the lemurs and New World monkeys had branched off.

Changes in our knowledge necessitate changes in how we assign names to species and groups of species, because scientific names are intended to not only be universal across languages, but to accurately reflect relatedness to other species and other groups. That has led to numerous name changes over the decades, as our means of classification have evolved.

Long before genetics was understood, we classified organisms on the basis of the obvious traits we could see, refining this over the years through details we learned about deep anatomy, behaviour and cell biology. The main hazard of relying on these things is the phenomenon of convergent evolution – whereby organisms that live in similar ways tend to evolve similar traits, even if the pathway towards building the traits is completely different. This can lead to incorrect presumptions. Convergent evolution has resulted, for

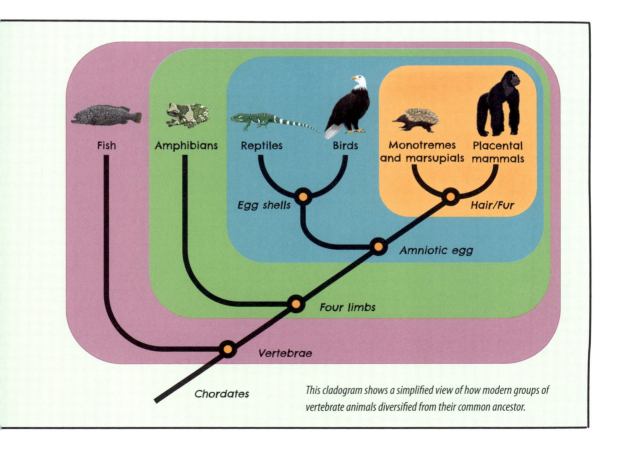

This cladogram shows a simplified view of how modern groups of vertebrate animals diversified from their common ancestor.

example, in mammals that resemble birds (bats) and fish (whales) – the idea of confusing them today might seem absurd, but texts of a couple of thousand years ago show that we did indeed consider bats to be birds and whales to be fish. Anatomy shows us that these animals' similarities are superficial, but much deeper-level convergently evolved similarities can be found, even in something as fundamental as the structure of cell components. Genetic coding is as fundamental as things can get, though (at least as far as we know today), and genetic evidence of relatedness overrides other biological evidence. Through lab analysis of genomes (including mitochondrial) we are building a much more accurate picture of the evolutionary history of life on Earth, and we can expect continued huge progress in this field of science.

CONSERVATION

As we saw in chapter 7, when breeding animals or propagating plants with a particular goal in mind, it's always been important to carefully keep track of who is related to who. In conservation of a species with a very small population, this is particularly important, to prevent the accumulation of

Black-footed ferret, an endangered species which may be saved from extinction by cloning technology.

problematic alleles and to maximize genetic variety, helping to make the species' population increasingly genetically robust through the generations. Now that it is possible to sequence individual genomes, the process of optimal breeding pair selection can be made more accurate.

Other applications of genetic science in conservation are those of cloning and gene editing (see chapter 7). Creating clones is more straightforward for some types of organisms than others. It is obviously possible to increase a small population through cloning, although natural breeding would usually be preferable. Since 2020, three cloned black-footed ferrets have been born, all using genetic material taken from a female ferret that died in the 1980s. Her

Toxic trickery

Cane toads were introduced from South America to Australia in the 1930s to control crop pests. As is often the way with such ideas, the toads themselves proved a much bigger problem, preying on native wildlife and devastating the populations of many native species. The northern quoll is a native predator, which can kill and eat cane toads despite the toads sometimes outweighing the quolls. However, the toads' flesh contains a neurotoxin, which makes them a deadly meal, and the marsupials have declined dramatically as the toads have increased. Conservationists have successfully trained captive-bred quolls to avoid toads, prior to releasing them in the wild. However, a more molecular approach is now underway – biologists at the University of Melbourne have been working on editing northern quoll DNA with genes from a species of predatory South American lizard. Because the lizard and cane toad evolved together, the lizard naturally evolved resistance to the toad's toxin, which the researchers hope can be passed on to the quoll to save it from extinction.

Northern quoll and cane toad.

tissues were preserved in the Frozen Zoo, a collection of preserved cellular material held at the Beckman Center for Conservation Research in San Diego. The first healthy clone proved unable to breed, for reasons likely to be unrelated to the cloning process, but hopes are high that her two sisters will be able to contribute new genetic variety to the black-footed ferret population.

MEDICINE

Although genetic science is progressing at high speed, the sheer volume of data that geneticists must process means that identifying which gene or genes are involved in various health conditions can be frustratingly slow, even if evidence of heredity shows clearly that a condition does have a genetic component. However, as the amount of research grows, the timescale is shrinking.

For some conditions (both heritable and not) where a mutated allele causes cells to manufacture a protein that works badly (or not at all), genetic engineering can help. Here, technicians can genetically engineer some of the patient's cells by replacing the faulty allele with one that codes for the correct protein, then infuse the relevant body tissues with a population of these correctly functioning cells. It may be necessary to maintain a lineage of the modified cells in the lab, so that the procedure can be repeated (as there will be original, badly functioning cells still present in the patient's body too), but this type of treatment is potentially less invasive and more effective than drug treatments. As we saw in chapter 9, it has already been used successfully to treat sickle cell disease. Many other diseases could be treated with cell-based gene therapies, and the technique also has the potential to help with wound-healing and cellular regeneration.

FOOD

From lab-grown meats to chickens' eggs that you can consume to cure an allergy to cat fur, the present (and future) impacts of genetic science on the food we can choose to eat is tremendous. Through selective breeding we have already taken numerous wild plant species and coaxed them into being super-charged versions of themselves – more productive with larger and more abundant fruit or grains, faster-growing, better-tasting, more resistant to disease and attack by crop-eating animals and to look much more appetizing. We have made our livestock similarly more productive and large-bodied, as well as moderating their temperaments so that they are relaxed in our company and are easier to handle. These processes have been a long time in the making, and careful selective breeding is still practised to keep a lineage progressing or holding steady along the right lines as the generations move on. Genetic engineering, however, allows for some serious fine-tuning.

Note: *The first burger made of lab-grown meat, unveiled to the world in 2013, cost $300,000 to produce. However, in just two years' time the developers had brought their unit cost down to just $11.36.*

Wheat is one of the most important food plants in the world – its grains are the source of flour for bread, pasta and noodles, while the rest of the plant provides food and bedding for domestic

Grow your own meat

Growing edible meat as living tissue in labs is, in theory, an ethical and economical way of producing this highly nutritious food. It would also be (again, in theory) possible to use gene-splicing techniques to boost the cultured meat's content of various nutrients, as well as modify its texture and flavour, although meat produced in this way is typically not genetically modified at present. In reality, consumers are even more cautious about cultured meat than they were about the first genetically modified plants on the market in the 1990s, and at present the process falls a long way short of being economically viable on a large scale. However, it certainly holds appeal for those who want to eat meat without there being any actual animals involved; and without the need to provide space and care for living animals, a well-designed facility could eventually produce a lot more meat for a much smaller footprint and workload than could be achieved in conventional farming.

The process of culturing edible meat in the lab.

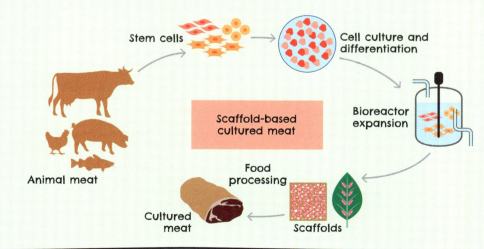

animals. The wheat genome was fully sequenced in 2018, and one of the edits currently in development (at the John Innes Centre in Norwich, UK) is a form of wheat grain that is higher in iron. Meat is rich in bioavailable iron, but if a wheat form could be developed that delivered a comparable amount of this essential nutrient, that would go a long way towards addressing the prevalence of anaemia (especially in young women and girls). The World Health Organization estimates that, worldwide, anaemia affects 40 per cent of children between 6 months and 5 years old, 37 per cent of pregnant women, and 30 per cent of women aged 15–49.

The food industry is a major cause of wildlife habitat loss and climate change worldwide, so development of plants that make more efficient use of land area and resources is an important focus for research. Teams at several universities are working on genetically modified strains of rice that will grow well in areas prone to drought (and so require less additional water), to photosynthesize more efficiently and to produce less methane.

COMPUTER SCIENCE

When we think of computer science and genetics, we acknowledge the crucial role computers have played in handling the colossal reams of data involved in genome sequencing, but what we know about genetics can also help us to improve computer processes. For example, the process of evolution by natural selection – on a genetically varied population with heritable traits – will, over time, produce organisms that are better adapted to their environment (or it will bring about their extinction). Modelling this process digitally has proved a useful and relatively low-effort way of building an artificial intelligence in computer science.

Unlike real-world evolution, a modelling simulation can be pre-programmed with an end goal, or simply allowed to run its course. Simulations such as GenePool, created by Jeffrey Ventrella, model evolution by natural selection. This is achieved through a population of virtual organisms with varied, genetically determined traits that must compete for food and breeding partners, and will pass a share of their genes on to each offspring. Over just a few generations, most of the initial diverse and random body forms have died out and are replaced with several distinct 'species' which are incredibly efficient at meeting the twin goals of grabbing food and grabbing mates. GenePool is a fun and educational tool, but genetic modelling has also been used for very serious purposes, such as, for example, to create the most effective and efficient design of antennae for satellites.

THE FUTURE OF GENETIC SCIENCE >>

The fruitful coming together of genetics and computer science is one of many signs that in this modern world we should perhaps no longer regard 'science' as a series of related but separate fields of study. After all, genetics touches upon many domains of science besides the biological, including social science. Our species has achieved extraordinary progression through everything we have learned about the workings of science, and learning for its own sake brings joy as well as progress, but our progression has also created problems – within ourselves as a species and throughout the wider world. Just as our genes drive us on to survive, so we need to use the fabulous intelligence that our genes have given us (along with the fabulous if slightly scary artificial intelligence that we have begun to develop) to take on the challenges of planning a sustainable future for our genes, ourselves and our world.

Neural networks work by passing data between interlinked nodes, which quickly produce outputs that can be compared, cross-referenced and combined.

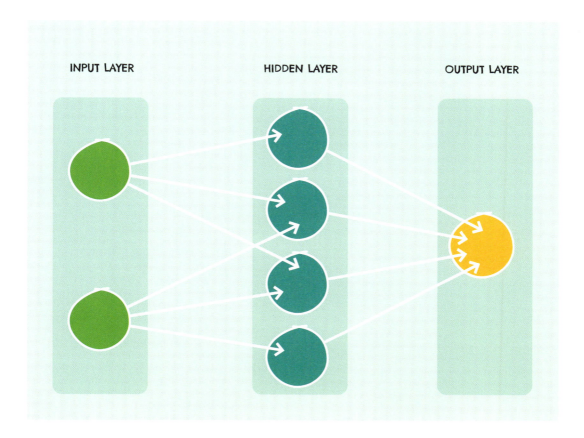

INDEX

A

alleles 43, 44, 45, 47–9, 50, 51, 53, 55, 56, 59–61, 64, 66, 71, 73, 74–5, 76, 78, 79, 87–8, 104, 105, 120–1

allosomes 21, 59, 67, 68, 102–3

artificial selection 83–8, 89

autosomes 21, 59, 101–2

B

bacterial chromosomes 15, 18

Barr body 55, 99

Barr, Murray 55

Belyaev, Dmitri 85

blood transfusions 108–10, 111

blood types 105–8, 110–11

Blundell, James 108

C

cell division 15, 22, 24, 27–8, 41–5, 43–4, 98

cells 15, 18

centromere 24, 41

Chase, Martha 14

chimeras 55–7

chromosomes 10

  anatomy of 22, 24–5

  bacterial 17

  karyograms 25–7

  pairs of 21–2

  ploidy sets 27–9

  staining of 22

cladistics 118–19

classification with DNA 77

clinal variation 75–6

cloning 121–2

co-dominance 62–3

codons 13

computer science 124–5

conservation 120–2

Crick, Francis 14, 15

cytogenetic bands 24–5, 30

D

Darwin, Charles 94

Dawkins, Richard 31, 33

diploid sets 27, 28, 102

DNA (deoxyribonucleic acid) 10

  in cells 15

  classification with 77

  description of 9–13

  discovery of 14

  double helix 16

  in forensic science 116–17

  junk DNA 18

  in natural world 15, 18

  replication of 43–5

  splicing 90–1

  testing 115–16

DNA databases 116

DNA testing 115–16

Dolly the sheep 88

domestication 83–8, 89

dominant/recessive alleles 59–60

double helix 16–17

E

endosymbiosis 19

enzymes 13, 37, 38, 41, 44, 50–1, 97, 98

epigenetics

  gene expression 95–7

  genetic memory of trauma 99

  nature/nurture debate 93–4

  reprogramming process 98

  X-inactivation 97–9

evolution

  clinal variation 75–6

  gene pool 71–2

  inbreeding 76, 78–9

  polymorphism 74–5

  sexual selection 72–4

  timeline of 79–81

F

food 122–4

forensic science 116–17

Franklin, Rosalind 14, 15

Frozen Zoo 121–2

G

gene editing 121

gene expression 95–7

gene pool 71–2

gene therapy 112–13

GenePool 124

genes

  alleles 47–9

  altruism and selfishness 31–2

  description of 6–7

  hox genes 31

  key vocabulary in 7

  mapping of 29–30

  and memes 33

genetic disorders 101–4

genetic engineering 88, 90, 122–4

genetic memory of trauma 99

genomes 10, 15, 18, 19, 21–2, 29–30, 48, 49, 50, 80–1, 83, 88, 96, 98, 112, 116, 120, 121, 124

genotypes 53, 55, 57, 60–3, 64, 107

Gosling, Ray 15

H

haploid sets 27–8

INDEX >> 127

Hershey, Alfred 14

heterogametic sex 21, 65

heterozygous genotype 61, 62–3, 65, 66, 104

hinnies 23

histones 22, 24, 96–7

homogametic sex 21, 63, 65, 67

homologues 43–4

homozygous genotype 61, 62, 63, 65–6, 104

hox genes 31

Human Genome project 29–30

I

incomplete dominance 63

inheritance

co-dominance 62–3

dominant/recessive alleles 59–60

genotypes 60–2

incomplete dominance 63

lethal genes 65–7

phenotypes 60–2

sex-linked traits 63–5

J

junk DNA 18

karyograms 25–7

L

Lamarck, Jean-Baptiste 94

Landsteiner, Karl 111

lethal genes 65–7

Lyon, Mary 55

M

mapping genones 29–30

medicines 122

memes 33

Mendel, Gregor 14, 59

messenger RNA (mRNA) 35, 37–9, 40, 52, 91, 97

methyl groups 12, 24, 96–7

methylation process 96–7

Miescher, Johann Friedrich 14

mitochondria 15, 18, 19, 28, 45

mitochondrial DNA 15, 18, 19, 29, 45, 115, 120

mitosis 15, 22, 24, 27–8, 41–5, 43–4, 98

monosomies 101–3

mules 23

mutations 45, 50–4, 86–8, 89

N

National DNA Database (NDNAD) 116

nature/nurture debate 93–4

nitrogenous base 10, 11, 12, 15

nuclear DNA 17, 19, 45

nuclear membrane 35, 39, 42

nucleic acids 10, 35

nucleotides 10–13, 14, 15, 16, 18, 35, 37, 39, 40, 41, 45, 52, 97

O

On the Origin of Species (Darwin) 94

organ donation 105

P

pentose sugar 10, 11, 16, 35

phenotypes 53, 60–2, 63, 65, 67, 87, 103, 107

Philosophie zoologique (Lamarck) 94

phosphate group 10, 11, 16, 35

ploidy sets 27–9

polymorphism 74–5

polyploid sets 28–9

population bottlenecks 76, 78

prokaryotes 17, 73, 79

protein synthesis 37–40

Punnet, Reginald 61

Punnett squares 60–1, 65

R

replication of DNA 43–5

reprogramming process 98

ribosomes 18, 35, 36, 37, 39–40, 41

RNA (ribonucleic acid) 6, 10, 11, 12, 29, 35–6, 37, 38, 39, 91, 97

S

screening 103

Scripps Research Institute 36

Selfish Gene, The (Dawkins) 31, 33

sex determination 67–9

sex-linked traits 63–5

sexual dimorphism 73–4

sexual selection 72–4

sister chromatids 24, 41–2, 43–4

splicing 38, 39, 90–1

stud books 7

T

T-cells 112–13

taxonomy 118–20

telomeres 24, 25

template strand 37, 41

tissue donation 105

transcription RNA (tRNA) 35, 37, 38, 40

trisomies 101–3

V

Ventrella, Jeffrey 124

W

Waldeyer-Hartz, Heinrich Wilhelm Gottfried von 22

Watson, James 14, 15

Wilkins, Maurice 15

World Health Organization 124

X

X-inactivation 53, 55, 56–7, 97–9

PICTURE CREDITS

Tobias Ambjoernsson 16

Geeks for Geeks 99

Getty 85

National Human Genome Research Institute 25, 26

Nature Picture Library 73

Paul Oakley 30

Researchgate.net 54, 68

Shutterstock 9, 10, 11, 11, 12, 14, 14, 14, 17, 18, 24, 24, 28, 32, 33, 35, 42, 43, 45, 47, 48, 48, 49, 50, 51, 55, 56, 57, 59, 62, 64, 66, 68, 68, 69, 72, 73, 77, 78, 78, 87, 87, 89, 90, 93, 94, 103, 111, 111, 115, 119, 120, 121, 121

Wikimedia Commons 22, 106

David Woodroffe 13, 19, 23, 27, 36 Bioninja.com, 37 National Human Genome Research Institute, 38 Wikimedia Commons, 40 Wikimedia Commons, 44, 57, 71, 75 Researchgate.net, 84, 91 Innovativegenomics.org, 97, 98 Researchgate.net, 101 Bioninja.com, 104 fcneurology.net, 113 Wikimedia Commons, 117 learn-genetics.b-cdn.net, 123 pub.mdpi-res.com/bioengineering/bioengineering